THE REAL ANALYSIS LIFESAVER

RAFFI GRINBERG

PRINCETON UNIVERSITY PRESS
Princeton and Oxford

Copyright © 2017 by Princeton University Press
Published by Princeton University Press, 41 William Street,
Princeton, New Jersey 08540
In the United Kingdom: Princeton University Press, 6 Oxford Street,
Woodstock, Oxfordshire OX20 1TR

press.princeton.edu

Jacket illustration by Dimitri Karetnikov
Jacket graphics courtesy of Shutterstock

All Rights Reserved

Library of Congress Cataloging-in-Publication Data

Names: Grinberg, Raffi, 1990–
Title: The real analysis lifesaver : all the tools you need to understand proofs / Raffi Grinberg.
Description: Princeton : Princeton University Press, 2016. | Series: A Princeton lifesaver study guide | Includes bibliographical references and index.
Identifiers: LCCN 2016020708| ISBN 9780691173870 (hardcover : alk. paper) | ISBN 9780691172934 (pbk. : alk. paper)
Subjects: LCSH: Mathematical analysis. | Functions of real variables. | Numbers, Real.
Classification: LCC QA299.8 .G75 2016 | DDC 515/.8—dc23 LC record available at https://lccn.loc.gov/2016020708

British Library Cataloging-in-Publication Data is available

This book has been composed in Times New Roman with Stencil and Avant Garde

Printed on acid-free paper. ∞

Typeset by Nova Techset Pvt Ltd, Bangalore, India
Printed in the United States of America

1 3 5 7 9 10 8 6 4 2

CONTENTS

Although I recommend that you read all of the chapters in order to cover a typical real analysis curriculum (and because the material builds on itself), it is possible to take a "fastest path" through this book using only the chapters marked with a *.

Preliminaries — 1
1. Introduction — 3
2. Basic Math and Logic* — 6
3. Set Theory* — 14

Real Numbers — 25
4. Least Upper Bounds* — 27
5. The Real Field* — 35
6. Complex Numbers and Euclidean Spaces — 46

Topology — 59
7. Bijections — 61
8. Countability — 68
9. Topological Definitions* — 79
10. Closed and Open Sets* — 90
11. Compact Sets* — 98
12. The Heine-Borel Theorem* — 108
13. Perfect and Connected Sets — 117

Sequences — 127
14. Convergence* — 129
15. Limits and Subsequences* — 138
16. Cauchy and Monotonic Sequences* — 148
17. Subsequential Limits — 157
18. Special Sequences — 166
19. Series* — 174

20. Conclusion — 183

Acknowledgments — 187

Bibliography — 189

Index — 191

PRELIMINARIES

CHAPTER 1

Introduction

Slow down there, hotshot. I know you're smart—you might have always been good with numbers, you might have aced calculus—but I want you to *slow down*. Real analysis is an entirely different animal from calculus or even linear algebra. Besides the fact that it's just plain harder, the way you learn real analysis is not by memorizing formulas or algorithms and plugging things in. Rather, you need to read and reread definitions and proofs until you understand the larger concepts at work, so you can apply those concepts in your own proofs. The best way to get good at this is to take your time; read slowly, write slowly, and think carefully.

What follows is a short introduction about why I wrote this book and how you should go about reading it.

Why I Wrote This Book

Real analysis is hard. This topic is probably your introduction to proof-based mathematics, which makes it even harder. But I very much believe that anyone can learn anything, as long as it is explained clearly enough.

I struggled with my first real analysis course. I constantly felt like I was my own teacher and wished there was someone who could explain things to me in a clear, linear fashion. The fact that I struggled—and eventually pulled through—makes me an excellent candidate to be your guide. I easily recall what it was like to see this stuff for the first time. I remember what confused me, what was never really clear, and what stumped me. In this book, I hope I can preempt most of your questions by giving you the explanations I would have most liked to have seen.

My course used the textbook *Principles of Mathematical Analysis*, 3rd edition, by Walter Rudin (also known as Baby Rudin, or That Grueling Little Blue Book). It is usually considered the classic, standard real analysis text. I appreciate Rudin now—his book is well organized and concise. But I can tell you that when I used it to learn the material for the first time, it was a *slog*. It never explains anything! Rudin lists definitions without giving examples and writes polished proofs without telling you how he came up with them.

Don't get me wrong: having to figure things out for yourself can be of tremendous value. Being challenged to understand why things work—without linear steps handed to you on a silver platter—makes you a better thinker and a better learner. But I believe that as a pedagogical technique, "throwing you in the deep end" without teaching you how to swim is only good in moderation. After all, your teachers want you to learn, not drown. I think Rudin can provide all the throwing, and this book can be a lifesaver when you need it.

I wrote this book because if you are an intelligent-but-not-a-genius student (like I was), who genuinely wants to learn real analysis... you need it.

What Is Real Analysis?

Real analysis is what mathematicians would call the *rigorous* version of calculus. Being "rigorous" means that every step we take and every formula we use must be proved. If we start from a set of basic assumptions, called *axioms* or *postulates*, we can always get to where we are now by taking one justified step after another.

In calculus, you might have proved some important results, but you also took many things for granted. What exactly *are* limits, and how do you really know when an infinite sum "converges" to one number? In an introductory real analysis course, you are reintroduced to concepts you've seen before—continuity, differentiability, and so on—but this time, their foundations will be clearly laid. And when you are done, you will have basically proven that calculus *works*.

Real analysis is typically the first course in a pure math curriculum, because it introduces you to the important ideas and methodologies of pure math in the context of material you are already familiar with.

Once you are able to be rigorous with familiar ideas, you can apply that way of thinking to unfamiliar territory. At the core of real analysis is the question: "how do we expand our intuition for certain concepts—such as sums—to work in the infinite cases?" Puzzles such as infinite sums cannot be properly understood without being rigorous. Thus, you must build your hard-core proving skills to apply them to these new (not-from-high-school-calculus), more interesting problems.

How to Read This Book

This book is not intended to be concise. Take a look at Chapter 7 as an example; I spend several pages covering what Rudin does in just two. The definitions are followed by examples in an attempt to make them less abstract. The proofs here are intended to show you not just *why* the theorem is true but also *how* you could go about proving it yourself. I try to state every fact being used in an argument, instead of omitting the more basic ones (as advanced mathematical literature would do).

If you are using Rudin, you'll find that I've purposely tried to cover all the definitions and theorems that he covers, mostly in the same order. There isn't a one-to-one mapping between this book and Rudin's (Chapter 7 math joke!); for example, the next chapter explains the basic theory of sets, whereas Rudin holds off on that until after covering real numbers. I also include a few extra pieces of information for your enrichment. But by following his structure and notations as closely as possible, you should be able to go back and forth between this book and his with ease.

Unlike some other math books—which are meant to be glanced at, skimmed, or just referenced—you should read this one linearly. The chapters here are deliberately short and should contain the equivalent of an easily digestible one-hour lecture. Start at the beginning of a chapter and don't jump around until you make it to the end.

Now for some advice: read actively. Fill in the blanks where I tell you to. (I purposely didn't include the answers to these; the temptation to peek would just be too great.) Make notes even where I don't tell you to. Copy definitions into your notebook if you learn by repetition; draw lots of figures if you learn visually. Write any questions you may have in the margins. If, after reading a chapter twice, you still have unanswered questions, ask your study group, ask your TA, ask your professor (or ask all three; the more times you hear something, the better you'll learn it). Within each chapter, try to summarize its main ideas or methods; you'll find that almost every topic has one or two tricks that are used to do most of the proofs.

If your time is limited or you are reviewing material you've already learned, you can use the following icons to guide your skimming:

- Here begins an example or a proof that is figured out step by step.

- This is an important clarification or thing to keep in mind.

- Try this fill-in-the-blank exercise!

- This is a more complicated topic that is only mentioned briefly.

Extra resources never hurt. In fact, the more textbooks you read, the better your chances of success in learning advanced mathematics. The best strategy is to have one or two primary textbooks (for example, this one with Rudin) whose material you are committed to learning. Complement those with a library of other books from which to get extra practice and to look up an explanation if your primaries are not satisfactory. If you choose to disregard this and try to learn *all* the material in *all* the real analysis books out there... good luck to you!

This book covers most of a typical first-semester real analysis course, though it's possible your school covers more material. If this book ends before your course does, don't panic! Everything builds on what comes before it, so the most important factor for success is an understanding of the fundamentals. We will cover those fundamentals in detail, to make sure you have a solid foundation with which to swim onward (while avoiding mixed metaphors, such as this one).

For a list of some recommended books, along with my comments and criticisms, see the Bibliography.

Once you turn the page, we'll begin learning by going over some basic mathematical and logical concepts; they are critical background material for a rigorous study of real analysis. (How many times have I used the word *rigorous* so far? This many: $\lim_{n\to\infty} \frac{n^\alpha}{(1+p)^n} + 7$.)

CHAPTER 2

Basic Math and Logic

If you've seen some of this stuff before, great! If not, don't worry—we'll take it nice and slow.

Some Notation

What follows are some notational conventions which you should become comfortable with.

The symbol $\forall$ stands for "for all" or "for every" and can also be read "as long as." For example, the definition of even numbers tells us: n is divisible by 2 $\forall n$ even. Read: "n is divisible by 2 for all even numbers n," or "n is divisible by 2 as long as n is even."

The symbol $\exists$ stands for "there exists" or "there is some." For example, one definition of the number e tells us: $\exists a$ such that $\frac{d}{dx} a^x = a^x$. That statement is true, since such a number a does exist; it is $e = 2.71828\ldots$

 Note that the following two statements have completely different meanings:

$$\forall x, \exists y \text{ such that } y > x$$

$$\exists y \text{ such that } y > x, \forall x$$

The first means that given any x, there is some y greater than it. The second means that there exists some y which is greater than *every* possible x. If x and y are real numbers, then the first statement is true, since for any x we can set $y = x + 1$. The second is false, since no matter how big a y we choose, there will always be another number bigger than it.

A *sequence* is a list of numbers, indexed in order by integers. For example, $2, 4, 6, \ldots$ is a sequence, and the $\ldots$ symbol "$\ldots$" indicates that it extends infinitely in a similar pattern. In the $2, 4, 6, \ldots$ example, the 10th element of the sequence is 20. By definition, a sequence continues on forever (so just the numbers $2, 4, 6$ is not a sequence).

Sequences can also be made up of variables, such as $x_1, x_2, x_3, \ldots$. We say that x_i is the ith element of the sequence, as long as i is a positive integer (so using the notation from above, x_i is the ith element of the sequence, $\forall i \geq 1$). The integer subscript of a particular x is the *index* of that element of the sequence.

The sum of elements in a pattern can be concisely expressed in *summation notation* using the Greek letter $\sum$ (capital sigma). For example, the sum of the first n integers can be written as $\sum_{i=1}^{n} i$, which is read: "the sum from $i = 1$ until $i = n$ of i." As you might have noticed, by convention, the index of the sum always takes on integer values, starting at the subscript of the sigma and ending at the superscript of the sigma. Another example is $\sum_{i=1}^{n} 1$, which is read: "The sum from $i = 1$ until $i = n$ of 1," which is just the sum of $1 + 1 + 1 + \ldots$, n times, which equals n.

Summations can also be written over a sequence (which, remember, is always infinite), and these are called *infinite series*, or just *series*. For example, the sum of all the elements in the aforementioned sequence 2, 4, 6, . . . , can be written as the series $\sum_{i=1}^{\infty} 2i = 2 + 4 + 6 + \ldots$. Another example is $\sum_{i=0}^{\infty} \frac{1}{i!}$, which actually equals the number e.

There are certain groups of numbers that have their own symbols:

- $\mathbb{N}$ is the set of all natural numbers. These are the positive integers, not including 0.
- $\mathbb{Z}$ is the set of all integers, including 0 and negative integers.
- $\mathbb{Q}$ is the set of all rational numbers These are defined as numbers of the form $\frac{m}{n}$, where $m \in \mathbb{Z}$ and $n \in \mathbb{Z}$.
- $\mathbb{R}$ is the set of all real numbers. We will define what *real numbers* actually are later.

You can remember these symbols by the following mnemonics: N is for Natural numbers, R is for Real numbers, Q is for Quotients, and Z is for integerZ.

Through Chapter 4, we assume all the usual facts that let us perform algebra on numbers in $\mathbb{N}$, $\mathbb{Z}$, and $\mathbb{Q}$. In Chapter 5, we'll take a closer look at these properties.

Formal Logic

What follows are some concepts from the study of logic, which we will use over and over again in proofs.

A logical statement is *equivalent* to another statement whenever it is only possible for them to be either both true or both false. For example, "I have been alive for 5 years" is equivalent to "I am 5 years old"—since if one is true, then so is the other; if one is false, then so is the other.

The symbol $\implies$ stands for "implies." For example, the following four statements are equivalent to each other:

Statement 1. If $n = 5$, then n is in $\mathbb{N}$.
Statement 2. n is in $\mathbb{N}$ if $n = 5$.
Statement 3. $n = 5$ only if n is in $\mathbb{N}$.
Statement 4. $n = 5 \implies n$ is in $\mathbb{N}$.

Note that these are *not* equivalent to "$n = 5$ if n is in $\mathbb{N}$." (Also, that statement is clearly not true, since there exist natural numbers that are not equal to 5—my personal favorite being 246,734.)

The symbol $\iff$ stands for "if and only if" (abbreviated *iff*), and it is used to state that both directions of an implication are true. For example, "n is even $\iff n$ is divisible by 2." The left statement implies the right statement, and vice versa. This particular *iff* statement is true, since it is the definition of even numbers.

Figure 2.1. The fact **A** is completely contained in **B**. If x is in **A**, then x is also in **B**.

There is a slightly confusing mathematical convention for writing definitions. Theoretically, all definitions should be written with "if and only if." For example, "a number is called even if and only if it is divisible by 2." The "if" goes both ways, since "even" is just a name we assign to certain numbers. However, mathematicians are lazy; to save time, they usually write definitions with just "if" instead of "if and only if." Don't be confused! If you see the following:

Definition *(Even)*
*A number is called **even** if it is divisible by 2.*

You should read:

Definition *(Even)*
*A number is called **even** if and only if it is divisible by 2.*

An arbitrary statement we might want to prove can be expressed as **A** $\implies$ **B**, where **A** and **B** are any facts.

The statement's *converse* is **B** $\implies$ **A**. Just because a statement **A** $\implies$ **B** is true, does not mean its converse **B** $\implies$ **A** is true. For example, we saw that $n = 5 \implies n$ is in $\mathbb{N}$, but n is in $\mathbb{N}$ does not imply $n = 5$.

The statement's *inverse* is $\neg$**A** $\implies$ $\neg$**B** (here, the symbol $\neg$ means "not"). Again, just because a statement **A** $\implies$ **B** is true, does not mean its inverse $\neg$**A** $\implies$ $\neg$**B** is true. For example, we saw that $n = 5 \implies n$ is in $\mathbb{N}$, but $n \neq 5$ does not imply n is not in $\mathbb{N}$ (because n could, for instance, be the number 246,734).

If **A** $\implies$ **B** is a statement, then $\neg$**B** $\implies$ $\neg$**A** is its *contrapositive*. The statement **A** $\implies$ **B** is actually always equivalent to the statement $\neg$**B** $\implies$ $\neg$**A**. If one of those statements is true, so is the other; if one of them is false, so is the other.

Why is every statement equivalent to its contrapositive? It helps to think of **A** $\implies$ **B** as saying "if x is in **A**, then x is in **B**." In that reading, we can represent **A** as a set that is completely contained in **B**.

Figure 2.1 helps us visualize: if x is not in **B**, then it certainly cannot be in **A**.

Last, note that if A is some property the number x can have, then the following two statements are equivalent:

Statement 1. $\neg\,(\forall x, x$ has property $A)$.
Statement 2. $\exists x$ such that $\neg\,(x$ has property $A)$.

The first statement says "it is not true that every x has property A," and the second statement says "there is some x such that x does not have property A." Read these two out loud, and it should be obvious why they are the same.

Similarly, the following two statements are also equivalent to each other:

Statement 3. $\neg\,(\exists x$ such that x has property $A)$.
Statement 4. $\forall x, \neg\,(x$ has property $A)$.

Try to read the statements out loud, translating all the symbols into English.

Proof Techniques

There are many different ways to prove a theorem; sometimes, more than one method will work. There are five main techniques used throughout this book:

1. Proof by counterexample.
2. Proof by contrapositive.
3. Proof by contradiction.
4. Proof by induction.
5. Direct proof, in two steps.

Proof by Counterexample. In some cases, a proof may just be one counterexample. How would you prove the fact that not every integer is even? If I say "every integer is even," you just need to find one example of an integer that is not even, for instance, the number 3, to prove me wrong. Proofs by counterexample work for any statement of the form "$\exists x$ such that x has property A," or "$\neg(\forall x, x$ has property $A)$." For the first, we just need to find one x that has property A; for the second, we just need to find one x that does not have property A.

Example 2.1. (Proof by Counterexample)
Let's try to prove the following statement: "Not every continuous function is differentiable." To do so, we just need one counterexample—any function that is continuous but not differentiable will do—for instance, $f(x) = |x|$.

Now we need to prove rigorously that $|x|$ is continuous and that it is not differentiable. You'll learn how to do so later on in your study of real analysis.

This example shows us that thinking up a counterexample is only half the work; the hard part is to prove rigorously that it indeed meets all the necessary conditions.

Proof by Contrapositive. As we understood earlier, $\neg\mathbf{B} \implies \neg\mathbf{A}$ is equivalent to $\mathbf{A} \implies \mathbf{B}$. So in order to prove $\mathbf{A} \implies \mathbf{B}$, we could alternatively assume $\mathbf{B}$ is false and show that $\mathbf{A}$ is also false.

Example 2.2. (Proof by Contrapositive)
Let's try to prove the following statement: *For any two numbers x and y, $x = y$ if and only if $\forall \epsilon > 0, |x - y| < \epsilon$.* This asserts that two numbers are equal if they are arbitrarily close (meaning we can choose an arbitrary distance ϵ, and they will be closer to each other than that distance). Since the statement has *iff,* the implication is bidirectional, and we must prove both directions.

1. $x = y \implies \forall \epsilon > 0, |x - y| < \epsilon$.
   Proving this direction is simple. Assume $x = y$. Then $x - y = 0$, so $|x - y| = 0$. Since any ϵ we choose must be greater than 0, we have $|x - y| = 0 < \epsilon$. Thus $\forall \epsilon > 0, |x - y| < \epsilon$.

2. $\forall \epsilon > 0, |x - y| < \epsilon \implies x = y$.
   This statement should make intuitive sense. It's saying that if the distance between x and y is less than every positive number, the distance between them must equal 0.
   To prove this direction, we'll use the contrapositive. In this case, the statement $\neg \mathbf{B} \implies \neg \mathbf{A}$ is
   $$x \neq y \implies \neg(\forall \epsilon > 0, |x - y| < \epsilon).$$
   Remember from our discussion of logic that we can simplify the right-hand side to
   $$x \neq y \implies \exists \epsilon > 0 \text{ such that } \neg(|x - y| < \epsilon).$$
   Well, if $x \neq y$, then x must equal y plus some number z, where $z \neq 0$. So $|x - y| = |z|$. The absolute value of any non-zero number is always positive, so if we let $\epsilon = |z|$, then $\epsilon > 0$ and $|x - y| = \epsilon$. We have shown that $\exists \epsilon > 0$ such that $\neg(|x - y| < \epsilon)$, so we're done!

Proof by Contradiction. Not to be confused with a proof by contrapositive, a proof by contradiction is something entirely different. Let's say we are trying to show that $\mathbf{A} \implies \mathbf{B}$. If we assume that $\mathbf{A}$ is true but $\mathbf{B}$ is false, then something should go horribly wrong; we should end up with a *contradiction*, something that violates a fundamental mathematical axiom or definition, such as "$0 = 1$" or "5.3 is an integer." When this happens, we have shown that if $\mathbf{A}$ is true, it is impossible for $\mathbf{B}$ *not* to be true—otherwise the definitions of math would break down.

Example 2.3. (Proof by Contradiction)
Let's try to prove the theorem "$\sqrt{2}$ is not a rational number." In this case, if we put the theorem into the form $\mathbf{A} \implies \mathbf{B}$, the statement $\mathbf{B}$ is "$\sqrt{2}$ is not a rational number." Notice that there really isn't any statement $\mathbf{A}$—since the theorem is claiming that it is not necessary for anything besides the usual mathematical axioms to be true, in order for $\mathbf{B}$ to be true. Thus proof by contrapositive won't work. How about a proof by contradiction? Assume that $\sqrt{2}$ *is* in $\mathbb{Q}$, and show that something goes horribly wrong.

Before getting started on the proof, here are some general facts about numbers that will come in handy.

Fact 1. Any rational number $\frac{m}{n}$ can be simplified so that m or n (or both) are not even. (If m and n are both even, we can just divide the top and bottom by 2 and obtain a more simplified version of the same rational number.)

Fact 2. If $a = 2b$ for some integers a and b, then a must be even, since it is divisible by 2.

Fact 3. If a number a is odd, then a^2 is also odd, since a^2 is an odd number added to itself an odd number of times.

Now we can start. If $\sqrt{2}$ is rational, then by Fact 1 it can be expressed as a simplified fraction, so there exist integers m and n (not both even) such that $\left(\frac{m}{n}\right)^2 = 2$. Then $m^2 = 2n^2$, so by Fact 2, m^2 must be even. By Fact 3, if m were odd, m^2 would also be odd, so m must be even.

We can express m as $2b$ for some number b, so $m^2 = (2b)^2 = 4b^2$, which implies that m^2 is divisible by 4. Then $2n^2$ is also divisible by 4, so n^2 is even. By Fact 3 again, n must also be even.

Wait! Fact 1 told us that if $\sqrt{2}$ is rational, we can express it as $\frac{m}{n}$, where m and n are not both even. But we just showed that they both *are* even! We have contradicted a basic axiom about fractions, so the only possible logical conclusion is that our main assumption—that $\sqrt{2}$ is rational—must be false.

By the way, the same methodology of this argument also works for proving that the square root of *any* prime number is not rational.

 A proof by contradiction is generally considered to be a last-resort method. In many cases, if you prove something by contradiction, you can apply the same key steps to easily prove the theorem directly. In the $\sqrt{2}$ example, that is not the case, but just be aware: proof by contradiction is a good way to start thinking about a problem, but always check to see if you can go further and prove it directly (for bonus mathematical etiquette points).

 Proof by Induction. Mathematical induction works the same way as dominoes: if we set them all up, and then knock over just the first one, they will all fall down. Induction works for any proof in which we need to prove an infinite number of cases (actually, it must be a *countably* infinite number of cases—you'll understand what this means in Chapter 8).

Let's say we're able to set up the dominoes by proving the following: *if we assume the theorem is true for case 1, then it is also true for case 2; if we assume the theorem is true for case 2, then it is also true for case 3;* and so on. This can by summarized by proving that *if the theorem is true for $n - 1$, then it is also true for n*. Now all we need to do is knock down the first domino by proving that the theorem is true for case 1. They all fall down, since our setup tells us that once case 1 is true, so is case 2; and now that case 2 is true, so is case 3; and so on.

Knocking the first domino down is easier, so we usually do it first (this step is called the *base case*). Then we assume the theorem is true for the $n - 1$ case (this assumption is called the *inductive hypothesis*) and show that it is also true for the n case (this step is called the *inductive step*).

Example 2.4. (Proof by Induction)
Let's try to find a formula for the sum of the first n natural numbers, $1 + 2 + 3 + \ldots + n$. Using our notation from the previous section, this sum is equivalent to $\sum_{i=1}^{n} i$. If you play with this long enough, you might stumble on the answer:

$$\sum_{i=1}^{n} i = \frac{n(n+1)}{2}.$$

Try plugging in a few values to convince yourself that the formula works. To prove it, we'll need to be more rigorous (a few examples isn't a proof, since they don't exclude the possibility that a counterexample exists). We need to show that this formula works for every possible choice of n, which can be any positive integer. Therefore, induction is probably the best technique.

1. *Base Case.* We just need to show that the formula holds for the case $n = 1$. Well, $\sum_{i=1}^{1} i = 1 = \frac{1(1+1)}{2}$, so the first step is done. Yay! (Base cases are usually a breeze.)

2. *Inductive Step.* The inductive hypothesis lets us assume that the formula is true for $n-1$, so we can assume that $\sum_{i=1}^{n-1} i = \frac{(n-1)(n)}{2}$. Using this assumption, we want to show that the formula holds for n, meaning $\sum_{i=1}^{n} i = \frac{n(n+1)}{2}$. By making a substitution and simplifying, we can write

$$\sum_{i=1}^{n} i = \left(\sum_{i=1}^{n-1} i\right) + n = \frac{(n-1)(n)}{2} + n = \frac{n^2 - n}{2} + \frac{2n}{2} = \frac{n^2 + n}{2} = \frac{n(n+1)}{2},$$

and that's it!

Although it seems almost too simple, remember, there's no magic involved. We didn't "bootstrap" the proof or use circular logic. We just used the inductive step to say, "if it works for 1, then it works for 2; if it works for 2, then it works for 3; and so on," and because the base case says, "it works for 1," we have thus proved it for every possible positive integer choice of n (in other words, for every n in $\mathbb{N}$).

Direct Proof in Two Steps. None of the tricks we have covered show how we can prove $\mathbf{A} \implies \mathbf{B}$ directly, by assuming $\mathbf{A}$ is true then taking logical steps to end up with $\mathbf{B}$.

Coming up with a direct proof requires you to play around for a while, until you figure out what the crux of the problem is and how to solve it. In many cases, the crux will involve finding some magic function or variable that makes everything fall into place. Unless you are writing a textbook, the reader of your proof does not care *how* you solved the crux, he or she just wants to see why the theorem is true. Once you have a good idea of what key steps you'll use to prove the theorem, the next step is to write it cleanly in a linear fashion.

Example 2.5. (Direct Proof in Two Steps)
Remember from calculus that we define the *limit* of a function as $\lim_{x \to p} f(x) = q$ if and only if for every $\epsilon > 0$, there is some $\delta > 0$ such that

$$|x - p| < \delta \implies |f(x) - q| < \epsilon.$$

You'll understand what this means in more detail when you study the topic of continuity. For now, though, let's just look at the following statement: "For $f(x) = 3x + 1$, $\lim_{x \to 2} f(x) = 7$." You know this should be true, since all polynomials are continuous—and there are probably multiple ways to prove this particular statement—but let's try to do a direct proof, using only the definition we just saw.

First we do the scratchwork to figure out the key steps. For every possible choice of $\epsilon > 0$, we need to find the right $\delta > 0$ so that

$$|x - 2| < \delta \implies |f(x) - 7| < \epsilon,$$

or equivalently

$$-\delta + 2 < x < \delta + 2 \implies -\epsilon < 3x - 6 < \epsilon.$$

Thus our δ needs to make

$$\frac{-\epsilon + 6}{3} < x < \frac{\epsilon + 6}{3},$$

so we just need

$$\delta + 2 = \frac{\epsilon + 6}{3} \implies \delta = \frac{\epsilon}{3}.$$

Now that we have found our magic δ, we can write up the proof concisely: For any $\epsilon > 0$, let $\delta = \frac{\epsilon}{3}$. Then $\delta > 0$, and

$$\begin{aligned} |x - p| < \delta &\implies |x - 2| < \frac{\epsilon}{3} \\ &\implies 2 - \frac{\epsilon}{3} < x < 2 + \frac{\epsilon}{3} \\ &\implies -\epsilon < 3x - 6 < \epsilon \\ &\implies |(3x + 1) - 7| < \epsilon \\ &\implies |f(x) - q| < \epsilon. \end{aligned}$$

Thus we have $\lim_{x \to 2} (3x + 1) = 7$.

One more hint about writing proofs: if you get stuck, look at the facts you haven't used yet. In the previous example, the only real "fact" available was the definition of a limit I gave you; but in later topics, you'll have a host of definitions and theorems to call on. Chances are, applying one you have forgotten will pull you out of the morass.

In the future, we will place the symbol $\square$, which signifies Q.E.D., at the end of every proof. It stands for the Latin *quod erat demonstrandum*, which basically means (and here I paraphrase liberally), "we have proved what we said we would prove."

And we're off! Now you know everything you'll need to start learning real analysis. As promised, we'll spend the next chapter learning about sets before we look at the real numbers.

CHAPTER 3

Set Theory

Before we dive into real analysis, a basic knowledge of sets (and how to manipulate them) will be useful. What are sets? Well, not all numbers are real numbers. In fact, not all "things" we wish to consider are numbers at all. *Sets* are a useful abstraction. They contain *elements*, which can be real numbers, imaginary numbers, dollars, people, beluga whales, and so on.

In this chapter, we'll go over the basic notation and theorems used to describe abstract sets. When you think of operations on numbers, addition, subtraction, multiplication, and division usually come to mind. For sets, however, the basic operations we will learn about are *union, intersection,* and *complement.*

Definition 3.1. *(Set)*
A **set** *is a collection of* **elements**. *A set with an infinite number of elements is called an* **infinite set**.

Example 3.2. (Sets)
Here are some examples of sets and their notation:

- $\{1, 2, 3\}$
  The set containing the numbers 1, 2, and 3. We write $1 \in \{1, 2, 3\}$ to mean that 1 is an element of the set.
- A
  The set named A.
- $A = \{1, 2, 3\}$
  The set named A which contains the numbers 1, 2, and 3.
- $\{a, b, c\}$
  The set containing the elements named a, b, and c. (These elements are not necessarily numbers.)
- $\{A, B, C\}$
  The set containing the elements named A, B, and C. In general, uppercase letters are used to denote sets, so this set might contain three other sets.
- $\mathbb{R}$
  The set containing all the real numbers. For example, $\pi \in \mathbb{R}$. This is an infinite set.

- $\{x \in \mathbb{R} \mid x < 3\}$
  This notation reads: "the set of all x in $\mathbb{R}$ such that x is less than 3," so this is the set of all real numbers less than 3. This is also an infinite set.
- $\{p \in A \mid p \neq 3\}$
  The set of all elements in A which are not equal to 3. As we will see later, this set could also be denoted as $A \setminus \{3\}$, the set A without the set comprising the single element 3.
- $\emptyset$
  This set holds a special place in every mathematician's heart. It is called the *empty set*, and it is just that: the set with no elements. Any set that is not the empty set is called *nonempty*.
- $|A| = 3$
  This notation means that the size of A is 3, so A contains 3 elements. Note that $|\{A\}|$ is not necessarily equal to $|A|$. In the first case, the size of the set containing only the set A is $|\{A\}| = 1$, whereas if A contains 100 elements, $|A| = 100$.

Definition 3.3. *(Indexed Family)*
*If for each $i \in I$ there corresponds a set A_i, then $\mathscr{A} = \{A_i \mid i \in I\}$ is the **indexed family** of sets A with **index set** I.*

*In some cases, we may write an indexed family of sets as $\{A_\alpha\}$ when we want to work with an arbitrary indexing set (which has elements α). This type of indexed family is also called a **collection** of sets.*

Example 3.4. (Indexed Families)
If $A_n = \{1, 2, n^2\}$ for all $n \in \mathbb{N}$, then $A_3 = \{1, 2, 9\}$. Thus if

$$\mathscr{A} = \{A_n \mid n \in \mathbb{N}, n \leq 10\},$$

then $\mathscr{A}$ is a set of sets, which looks like

$$\mathscr{A} = \{\{1, 2, 1\}, \{1, 2, 4\}, \ldots, \{1, 2, 100\}\}.$$

Note that an indexed family doesn't need to be finite. In this example, if

$$\mathscr{B} = \{A_n \mid n \in \mathbb{N}\},$$

then $\mathscr{B}$ is an infinite set of sets, which looks like

$$\mathscr{B} = \{\{1, 2, 1\}, \{1, 2, 4\}, \{1, 2, 9\}, \ldots\}.$$

Don't be confused by the set $\{1, 2, 1\}$. It's the same as the set $\{1, 2\}$—it's just a more redundant way of writing it. For example, if we had the set $\{1\}$, you could actually write it as $\{1, 1, 1, 1\}$.
...But please don't.

Definition 3.5. *(Subset)*
*A set A is a **subset** of the set B, written $A \subseteq B$ or $B \supseteq A$, if every element of A is also an element of B. In this case, B is called a **superset** of A.*

A set A is **equal** to the set B, written $A = B$, if A and B have exactly the same elements. (If two sets are equal, they are the same set.) This is the same as saying that "both $A \subseteq B$ and $B \subseteq A$."

A set A is a **proper subset** of the set B, written $A \subset B$ or $B \supset A$, if every element of A is also an element of B, but B has some other elements outside of A. In symbols:

$$A \subset B \iff A \subseteq B \text{ and } A \neq B.$$

Example 3.6. (Subsets)
If $A = \{1, 2\}$ and $B = \{1, 2, 3\}$, then $A \subseteq B$. Furthermore, $A \subset B$.

Using our notation from Example 3.2, we can write $\{x \mid x \text{ is real}\} = \mathbb{R}$. Also, since every rational number is a real number, we can write $\mathbb{Q} \subseteq \mathbb{R}$. Furthermore, $\mathbb{Q} \subset \mathbb{R}$, since we know that there are some real numbers that are not rational (such as $\sqrt{2}$).

Weirdly imprecise mathematical convention. For some reason, the conventional notation is to write $A \subset B$ instead of $A \subseteq B$—even when A might be equal to B. I know this contradicts what I wrote earlier; it's just a convention I adopt so you can get used to the way most mathematicians write it. So when you see $A \subset B$, do not assume that A is a proper subset of B unless I explicitly say so.

Just to confuse you further. Note that there is a difference between writing $A \in B$ and $A \subset B$. The first means that B is a set of sets and the set A is one of its elements, whereas the latter means that B contains all the elements of A.

For example, take $A = \{1, 2\}$. To write $A \in B$, B would have to be something like $\{\{1, 2\}, \{3, 4\}, \{1\}, \{100\}\}$ (so B is a set of sets). If $A \subset B$, then B would have to be something like $\{1, 2, 3, 100\}$.

Theorem 3.7. *(Every Set Contains the Empty Set)*
For any set A, we have $\emptyset \subset A$.

Note that we cannot say that "$\emptyset$ is a proper subset of A for all sets A," since A might equal the empty set.

Proof. We must show that every element of $\emptyset$ is also an element of A. But $\emptyset$ has no elements, so all of its elements are in A. □

Definition 3.8. *(Intervals)*
The **open interval** (a, b) is the set

$$(a, b) = \{x \in \mathbb{R} \mid a < x < b\}.$$

The **closed interval** $[a, b]$ is the set

$$[a, b] = \{x \in \mathbb{R} \mid a \leq x \leq b\}.$$

The **half-open interval** is the set

$$[a, b) = \{x \in \mathbb{R} \mid a \leq x < b\},$$

or the set

$$(a, b] = \{x \in \mathbb{R} \mid a < x \leq b\}.$$

Some people also call open intervals *segments* and closed intervals (just plain) *intervals*, but this terminology is a bit old-fashioned (and more confusing). We'll stick with the terms *open interval* (a, b) and *closed interval* $[a, b]$ for consistency.

Example 3.9. (Intervals)
$(-3, 3)$ is an open interval and $[0, 99.5]$ is a closed interval. Both are subsets of the real numbers.

Definition 3.10. *(Union and Intersection)*
The **union** *of two sets A and B is the set comprising all elements of A together with all elements of B.*
In symbols, the union of A and B is:

$$A \cup B = \{x \mid x \in A \text{ or } x \in B\}.$$

*The **intersection** of two sets A and B is the set comprising all elements that are both in A and in B.*
In symbols, the intersection of A and B is:

$$A \cap B = \{x \mid x \in A \text{ and } x \in B\}.$$

*If $A \cap B = \emptyset$, then A and B are **disjoint**. Otherwise, we say that A and B **intersect**.*

Example 3.11. (Unions and Intersections)
The union $\{1, 2, 3\} \cup \{3, 4, 5\}$ is $\{1, 2, 3, 4, 5\}$, which we can also write as $\{n \in \mathbb{N} \mid 1 \leq n \leq 5\}$. The intersection $\{1, 2, 3\} \cap \{3, 4, 5\}$ is the set $\{3\}$, which consists of the single element 3.

Because intervals are sets, we can take their unions and intersections. For example, $[1, 3] \cup [2, 4] = [1, 4]$, and $[1, 3] \cap [2, 4] = [2, 3]$. Note that $(1, 3) \cup (3, 5) \neq (1, 5)$. Rather,

$$(1, 3) \cup (3, 5) = \{x \in \mathbb{R} \mid 1 < x < 3 \text{ or } 3 < x < 5\},$$

so $(1, 3) \cup (3, 5)$ is actually the simplest way to write that union. On the other hand, $(1, 3) \cap (3, 5) = \emptyset$.

Remember that by definition, intervals only contain real numbers. When we want to describe all the rational numbers between a and b, we write $(a, b) \cap \mathbb{Q}$. Note that $\mathbb{Q} \cup \mathbb{R} = \mathbb{R}$, and $\mathbb{Q} \cap \mathbb{R} = \mathbb{Q}$.

As a final example, here are some simple facts about unions and intersections.

> *Fact 1.* For any set A,
>
> $$A \cup A = A = A \cap A$$

> *Fact 2.* If x is in A or x is in B, then we can also say x is in B or x is in A (we just switched the order), which makes the following laws of commutativity and associativity trivial:
>
> $$A \cup B = B \cup A \text{ and } (A \cup B) \cup C = A \cup (B \cup C),$$

and

$$A \cap B = B \cap A \text{ and } (A \cap B) \cap C = A \cap (B \cap C).$$

Fact 3. $A \cup \emptyset = A$, and $A \cap \emptyset = \emptyset$. Thus every set is disjoint from the empty set.

Theorem 3.12. *(Properties of Unions and Intersections)*
The following properties hold for any sets A and B.

Property 1. $A \subset (A \cup B)$, and $A \supset (A \cap B)$.
Property 2. $A \subset B$ if and only if $A \cup B = B$.
Property 3. $A \subset B$ if and only if $A \cap B = A$.
Property 4. For any $n \in \mathbb{N}$,

$$A \cup (B_1 \cap B_2 \cap \ldots \cap B_n) = (A \cup B_1) \cap (A \cup B_2) \cap \ldots \cap (A \cup B_n)$$

Property 5. For any $n \in \mathbb{N}$,

$$A \cap (B_1 \cup B_2 \cup \ldots \cup B_n) = (A \cap B_1) \cup (A \cap B_2) \cup \ldots \cup (A \cap B_n)$$

Proof. These proofs are pretty straightforward; try filling in the blanks for Properties 3 and 5.

1. Let $x \in A$. Then $x \in A$ or $x \in B$.
   For the second half, let $x \in A \cap B$. Then $x \in A$ and $x \in B$, so $x \in A$.
2. Assume $A \subset B$. If we can show that $B \subset (A \cup B)$ and $B \supset (A \cup B)$, then we have shown $A \cup B = B$. The first fact, $B \subset (A \cup B)$, is true by that first property we just proved. We assumed that $A \subset B$, so we can just union both sides with B to get $(A \cup B) \subset (B \cup B) = B$, and we are done.
   To prove the converse, assume $A \cup B = B$. By the first property again, $A \subset (A \cup B)$, so we have $A \subset (A \cup B) = B$, so $A \subset B$.
3. This proof is almost the same as the last one.

BOX 3.1

> PROVING PROPERTY 3 OF THEOREM 3.12
>
> Assume $A \subset B$. We want to show $A \subset (A \cap B)$ and _____. The first fact is true: because $A \subset B$, if $x \in A$ then _____, so if $x \in A$ then $x \in A$ and $x \in B$. The second fact is true by _____.
>   Conversely, assume _____. By the first property, $B \supset$ _____, so we have $A =$ _____ $\subset B$.

4. Here's some notation to abbreviate what we're trying to prove:

$$A \cup \left(\bigcap_{i=1}^{n} B_i \right) = \bigcap_{i=1}^{n} (A \cup B_i).$$

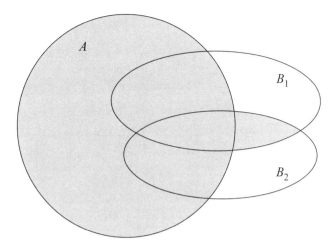

Figure 3.1. The shaded region is $A \cup (B_1 \cap B_2) = (A \cup B_1) \cap (A \cup B_2)$.

The big intersection sign works the same way as the summation notation $\sum_{i=1}^{n}$. Let's start by showing

$$A \cup \left(\bigcap_{i=1}^{n} B_i\right) \subset \bigcap_{i=1}^{n} (A \cup B_i),$$

then later prove

$$A \cup \left(\bigcap_{i=1}^{n} B_i\right) \supset \bigcap_{i=1}^{n} (A \cup B_i).$$

See Figure 3.1 for an illustration of this property. It's often helpful if you draw similar Venn diagrams when you're trying to understand ideas related to unions and intersections. Make sure your drawings account for every possible intersection—if, for example, you drew another blob called B_3 that looks like it only intersects with A (but not B_1 and B_2), you might end up with false conclusions about these sets.

Let $x \in A \cup (\bigcap_{i=1}^{n} B_i)$, so either $x \in A$ or $x \in \bigcap_{i=1}^{n} B_i$. If $x \in A$, then $x \in A \cup B_i$ for every i (since $A \subset (A \cup B_i)$, by the first property), so $x \in \bigcap_{i=1}^{n}(A \cup B_i)$. If $x \in \bigcap_{i=1}^{n} B_i$, then $x \in B_i$ for every i, so $x \in A \cup B_i$ for every i, so $x \in \bigcap_{i=1}^{n}(A \cup B_i)$.

To get the other direction of the subset equality, guess what we do? Pretty much the same thing. Let $x \in \bigcap_{i=1}^{n}(A \cup B_i)$, so $x \in A \cup B_i$ for every i. We now break it down into two possible cases:

> *Case 1.* $x \in A$. Clearly $x \in A \cup (\bigcap_{i=1}^{n} B_i)$. (Why? Because $A \subset A \cup (\bigcap_{i=1}^{n} B_i)$. See how we keep using that first property?)
>
> *Case 2.* $x \notin A$. Then x must belong to B_i for every i (since x is in A or B_i for every i). Thus $x \in \bigcap_{i=1}^{n} B_i$, so $x \in A \cup (\bigcap_{i=1}^{n} B_i)$.

5. These arguments aren't nearly as painful as they look. You just go for the $A \subset B$ and $A \supset B \implies A = B$ trick; for each half of the proof, take an element in the set and see where it naturally leads you. I've shortened this one down to symbols and arrows, so you can fill it in quickly. But first, draw a picture like Figure 3.1 so you can intuit why it should be true.

BOX 3.2

PROVING PROPERTY 5 OF THEOREM 3.12

$A \cap (\bigcup_{i=1}^{n} B_i) \subset \bigcup_{i=1}^{n} (A \cap B_i)$, since:

$x \in A \cap \left(\bigcup_{i=1}^{n} B_i \right) \implies$ _____ and _____

$\implies x \in$ _____ (for some i)

$\implies x \in A \cap B_i$ (for _____)

$\implies$ _____ or $x \in A \cap B_2$ or ... or _____

$\implies x \in \bigcup_{i=1}^{n} (A \cap B_i)$.

And _____, since:

$x \in \bigcup_{i=1}^{n} (A \cap B_i) \implies$ _____ (for some i)

$\implies$ _____ and _____ (for some i)

$\implies$ _____ or _____ or ... or _____

$\implies x \in A$ and $x \in \bigcup_{i=1}^{n} B_i$

$\implies$ _____.

□

Definition 3.13. *(Union and Intersection in Families)*
We define $\bigcup_{i \in I} A_i$, the **union** of the sets in the indexed family $\mathscr{A} = \{A_i \mid i \in I\}$, as the set of all elements which belong to A_i for at least one $i \in I$.

We define $\bigcap_{i \in I} A_i$, the **intersection** of the sets in the indexed family $\mathscr{A} = \{A_i \mid i \in I\}$, as the set of all elements that belong to A_i for every $i \in I$.

Note that this notation allows us to work with infinite unions and infinite intersections. Properties 4 and 5 of Theorem 3.12 apply to infinite unions and infinite intersections as well; in no part of the argument did we need to assume that the index i belongs to a finite index set.

Example 3.14. *(Unions and Intersections in Families)*
In terms of notation, note that $\bigcup_{n=1}^{\infty} A_n = \bigcup_{n \in \mathbb{N}} A_n$, and $\bigcap_{n=1}^{\infty} A_n = \bigcap_{n \in \mathbb{N}} A_n$. Of course, the infinite union of all natural numbers is $\bigcup_{n=1}^{\infty} \{n\} = \mathbb{N}$.

For any α, let $A_\alpha = \{\alpha\}$. Then $\bigcup_{\alpha \in \mathbb{R}} A_\alpha = \mathbb{R}$, and $\bigcap_{\alpha \in \mathbb{R}} A_\alpha = \emptyset$. Actually, for any indexing set I, we have $\bigcup_{\alpha \in I} \{\alpha\} = I$, and $\bigcap_{\alpha \in I} \{\alpha\} = \emptyset$.

For any $n \in \mathbb{N}$, let $A_n = (-\frac{1}{n}, \frac{1}{n})$. Then $\bigcup_{n=1}^{\infty} A_n = (-1, 1)$. Why? For each $m > n$, we have $A_m \subset A_n$, so by Property 3 of Theorem 3.12, the union is just the biggest set in the indexed family, which is the A with the smallest subscript. So

$$\bigcup_{n=1}^{\infty} A_n = A_1 = (-1, 1).$$

For any $n \in \mathbb{N}$, let $A_n = [0, 2 - \frac{1}{n}]$. Then $\bigcup_{n=1}^{\infty} A_n = [0, 2)$. (Note that this is a half-open interval. It does not contain the number 2, since there is no $n \in \mathbb{N}$ such that $2 \in A_n$.) Why? We need to wait until Theorem 5.5 to see why this is true. Also, $\bigcap_{n=1}^{\infty} A_n = [0, 1]$. Why? For each $m > n$, $A_m \supset A_n$, so by Property 3 of Theorem 3.12, the intersection is just the smallest set in the indexed family, which is the A with the smallest subscript. Thus

$$\bigcap_{n=1}^{\infty} A_n = A_1 = [0, 1].$$

Definition 3.15. *(Complement)*
The **complement** of a set A in another set B, written A^C in B or $B \setminus A$, is the set of elements that are in B but not A.

In symbols, the complement of A in B is:

$$B \setminus A = \{x \in B \mid x \notin A\}.$$

Example 3.16. *(Complements)*
The complement of $[-3, 3]$ in $\mathbb{R}$ is $(-\infty, -3) \cup (3, \infty)$, and the complement of $(-3, 3)$ in $\mathbb{R}$ is $(-\infty, -3] \cup [3, \infty)$. Note that $\mathbb{Q}^C$ in $\mathbb{R}$ is the set of all irrational numbers.

For any sets A and B, we have $(A \cup B) \setminus B = A \setminus B$, and $(A \cap B) \setminus B = \emptyset$. If $A \subset B$, then $A \setminus B = \emptyset$. (The $(A \cap B) \setminus B = \emptyset$ example is an instance of this one, because $(A \cap B) \subset B$.)

The complement of the complement of A is everything not in what's not in A, which is just A itself. So we always have $B \setminus (B \setminus A) = A$.

We can also use complements to describe sets without some of their elements. For example, if $A = \{1, 2, 100\}$, then $A \setminus \{100\}$ is the complement of $\{100\}$ in A, meaning everything in A except for the element 100, which is the set $\{1, 2\}$.

The next theorem is a standard result that is applied in almost every area of math. It is named after Augustus De Morgan, who formalized its cognate in the field of logic: for two statements **A** and **B**,

$$\neg(\mathbf{A} \text{ or } \mathbf{B}) \iff (\neg \mathbf{A}) \text{ and } (\neg \mathbf{B})$$

$$\neg(\mathbf{A} \text{ and } \mathbf{B}) \iff (\neg \mathbf{A}) \text{ or } (\neg \mathbf{B}).$$

Theorem 3.17. *(De Morgan's Law)*
Let E_α be any collection of sets (finite or infinite), and let them all be subsets of the set X. (In this theorem and the proof that follows, we will save space by writing "E_α^C" instead of the full "E_α^C in X.")

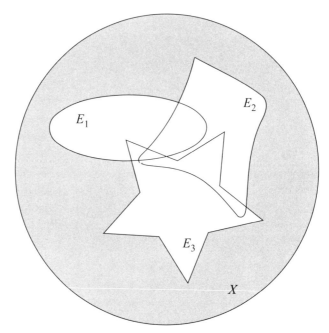

Figure 3.2. The shaded region is $(E_1 \cup E_2 \cup E_3)^C = E_1^C \cap E_2^C \cap E_3^C$.

Then the following statements hold:

$$\left(\bigcup_\alpha E_\alpha\right)^C = \bigcap_\alpha (E_\alpha^C),$$

$$\left(\bigcap_\alpha E_\alpha\right)^C = \bigcup_\alpha (E_\alpha^C).$$

See how these two statements are similar to the two logic results presented before? A union corresponds to the *or* logical operator, and an intersection corresponds to the *and* logical operator. Then De Morgan's law basically says that the negation of a union is the same as the intersection of negations and vice versa. Take a look at Figure 3.2 to see why the law makes sense.

Proof. How would *you* go about proving this? Maybe you would walk around the room to refresh your mind, get distracted by the TV, realize you're too tired for this so you go get a cup of coffee, run into someone you met at the beginning of the year whose name you've forgotten (so you give an awkward half-wave), sprint back to your room to see if getting exercise will help, realize that was a terrible idea since now you're even more tired *and* all sweaty, then sit down and stare and stare and stare... then get an epiphany: "Hey! I can just do what this author punk did for the last billion proofs!" That's right: just let $A = (\bigcup_\alpha E_\alpha)^C$ and let $B = \bigcap_\alpha (E_\alpha^C)$, then show $A \subset B$ and $B \subset A$.

Let $x \in A$, so by the definition of complement, $x \in X$ and $x \notin \bigcup_\alpha E_\alpha$. Then $x \notin E_\alpha$ for any α, so $x \in E_\alpha^C$ for every α. Thus $x \in \bigcap_\alpha (E_\alpha^C) = B$, so $A \subset B$.

Now let $x \in B$, so $x \in E_\alpha^C$ for every α, so $x \in X$ and $x \notin E_\alpha$ for any α. Then $x \notin \bigcup_\alpha E_\alpha$, so $x \in (\bigcup_\alpha E_\alpha)^C = A$, so $B \subset A$.

Now, to get the second statement, we just take complements. Because $\{E_\alpha\}$ are arbitrary sets, we can apply the statement we proved to the indexed family $\{E_\alpha^C\}$ to get

$$\left(\bigcup_\alpha E_\alpha^C\right)^C = \bigcap_\alpha (E_\alpha^C)^C = \bigcap_\alpha E_\alpha.$$

Take the complement of both sides to get what we wanted, namely, $\bigcup_\alpha E_\alpha^C = \left(\bigcap_\alpha E_\alpha\right)^C$.

When you look at a straightforward but dense proof like this, it's easy to let your eyes move forward while nodding your head like "yeah, that makes sense"—without really understanding what's going on. Just to make sure you're paying attention (and that you've truly become a set master), try copying this proof into your notebook. Then try copying it again onto another page, this time without looking here.

...Done yet? Okay, now you can go to sleep. □

Starting soon, we'll begin our study of real analysis with _____ numbers (fill in the blank!). The real numbers are actually an example of an ordered field, which is a specific type of set whose properties we'll explore in Chapter 5. But first, we'll learn everything there is to know about bounds.

REAL NUMBERS

CHAPTER 4

Least Upper Bounds

The natural numbers are the numbers you can count; the integers are the naturals with their negatives and zero; the rationals are the integers with their quotients; the reals are.... Actually, what are real numbers exactly? They're a combination of rational and irrational numbers, sure, but irrational just means "not rational"—we can't define what those numbers really are.

As we'll see in the next theorem, the fact that $\sqrt{2}$ is missing from $\mathbb{Q}$ puts a "hole" in the rational numbers, forming subsets with no smallest number and subsets with no largest number. These holes prevent $\mathbb{Q}$ from having something special called the *least upper bound property*.

To fill in these holes, we'll form $\mathbb{R}$, a superset of $\mathbb{Q}$ that does have the least upper bound property.

Theorem 4.1. *($\mathbb{Q}$ Has Holes)*
There exist subsets of the rational numbers that have no smallest number, and there exist others that have no largest number.

Proof. For this theorem, a proof by example will suffice—we just need to find one subset of $\mathbb{Q}$ that has no smallest number, and another with no largest number. (If you want to be technical, you'll notice the theorem says "subsets," plural, so we should actually find *two* examples of each type of subset. If this really bothers you, how about you find another example of each, as an exercise. Ha! That's what you get for being technical.)

By Example 2.3, we know $\sqrt{2} \notin \mathbb{Q}$. Let A be the set of all $p \in \mathbb{Q}$ such that $p^2 < 2$ and $p > 0$. To show that A has no single largest number, we must show that for any element of A, there is a greater element also in A. In symbols, we must show:

$$p \in A \implies \exists q \in A \quad \text{such that } q > p.$$

After fiddling around for a while, we might find that $q = \frac{2p+2}{p+2}$ works. (In the proof of Theorem 5.8, we'll come up with a general formula for finding numbers like q.)

We have

$$q = \frac{2p+2}{p+2}$$
$$= \frac{2p+2+p^2-p^2}{p+2}$$
$$= \frac{p(p+2)-p^2+2}{p+2}$$
$$= p - \frac{p^2-2}{p+2}.$$

Remember that $p^2 < 2$, so $p^2 - 2 < 0$; and $p > 0$, so $p + 2 > 0$. Thus $\frac{p^2-2}{p+2} < 0$, so $p - \frac{p^2-2}{p+2} > p$. In other words, $q > p$.

We still need to show $q^2 < 2$ (so that $q \in A$). Well,

$$q^2 - 2 = \frac{(2p+2)^2}{(p+2)^2} - 2$$
$$= \frac{(4p^2+8p+4)-2(p^2+4p+4)}{(p+2)^2}$$
$$= \frac{2(p^2-2)}{(p+2)^2}.$$

Again, we use $p^2 < 2$, so $2(p^2 - 2) < 0$; and $p > 0$, so $(p+2)^2 > 0$. Thus $\frac{2(p^2-2)}{(p+2)^2} < 0$, so indeed $q^2 - 2 < 0$.

To find a subset of $\mathbb{Q}$ with no smallest number, let B be the set of all $p \in \mathbb{Q}$ such that $p^2 > 2$ and $p > 0$. Here we need to show $p \in B \implies \exists q \in B$ such that $q < p$. It turns out that the same q as above works for B! Fill in the blanks to show that $q < p$ and that $q^2 > 2$.

BOX 4.1

> PROVING THE SUBSET B OF $\mathbb{Q}$ HAS NO SMALLEST NUMBER
>
> By the work from before, we know that
>
> $$q = p - \underline{}.$$
>
> Since $p^2 > 2$ for every $p \in$ _____, we know $2(p^2 - 2)$ _____ 0; and since $p > 0$, we know $p + 2 > 0$. Thus q equals p minus some positive number, so $q < p$.
>
> We also know that
>
> $$q^2 - 2 = \underline{}.$$
>
> Since $p^2 > 2$, we know _____ > 0; since $p > 0$, we know _____ > 0. Thus $q^2 - 2 >$ _____.

Later on, we'll want to refer back to these sets A and B. We said

$$A = \{p \in \mathbb{Q} \mid p^2 < 2 \text{ and } p > 0\},$$

$$B = \{p \in \mathbb{Q} \mid p^2 > 2 \text{ and } p > 0\}.$$

In other words, we can write

$$A = (0, \sqrt{2}) \cap \mathbb{Q},$$

$$B = (\sqrt{2}, \infty) \cap \mathbb{Q}.$$

To prove the previous theorem, we took for granted that $\mathbb{Q}$'s elements can be put in an order, with every rational number sandwiched between two other rational numbers. This property makes $\mathbb{Q}$ an *ordered set,* which we should define more formally.

Definition 4.2. *(Ordered Set)*
*An **order** on a set S is a relation, symbolized by $<$, that has the following properties:*

Property 1. For $x, y \in S$, exactly one of the following is true:

$$x < y, \text{ or } x = y, \text{ or } y < x.$$

Property 2. For $x, y, z \in S$, if $x < y$ and $y < z$, then $x < z$.

*If an order is defined on S, then S is an **ordered set**.*

The statement $x < y$ can also be written as $y > x$. The statement $x < y$ or $x = y$ can also be written $x \leq y$. In symbols, this means:

$$x \leq y \iff \neg(x > y),$$

and

$$x \geq y \iff \neg(x < y).$$

If a set is ordered, we can have a meaningful conception of minima and maxima.

Definition 4.3. *(Minimum and Maximum)*
*The **minimum** of an ordered set A is the smallest element in A. The **maximum** of an ordered set A is the largest element in A.*

Example 4.4. (Minima and Maxima)
We usually write $\min A$ or $\max A$. For example, $\min\{1, 2, 100\} = 1$ and $\max\{1, 2, 100\} = 100$.

Note that if $A \subset B$ and A has a minimum, then $\min A \geq \min B$. Why? Since $|A| \leq |B|$, the smallest element b of B may or may not be in A. If $b \in A$, then b is also the smallest element of A (since it is smaller than everything in B, which includes everything in A), so $\min A = \min B$; if $b \notin A$, then the smallest element of A must be something bigger than b (otherwise it would be the minimum of B). Similarly, if $A \subset B$, then $\max A \leq \max B$.

We can also take the minimum of the sizes of sets in an indexed family, meaning $\min_\alpha |A_\alpha|$. Consider the size of each $|A_\alpha|$ as an element of a set. Since each size is a number, that set of sizes is ordered, and we can take its minimum over all possible indexes α. For example, given any $\alpha \in \mathbb{R}$, let

$$A_\alpha = \{n \in \mathbb{N} \mid n \leq \alpha\}.$$

Then $\min_\alpha |A_\alpha| = 0$, since there is an $\alpha \in \mathbb{R}$ with no natural numbers smaller than it. For example, $A_{0.5} = \emptyset$ so $|A_{0.5}| = 0$.

What about $\max_\alpha |A_\alpha|$? For any $\alpha \in \mathbb{R}$, $A_{\alpha+1}$ has at least one more element than A_α. Thus given any potential maximum of $|A_\alpha|$, we can find some element greater than it. When this happens, we say that the maximum does not exist.

There are more cases in which we cannot take the minimum or maximum of an infinite set. For example, $\min(-3, 3)$ is undefined, because there is no smallest number in that interval. (If you think you've got a smallest number a, we can always find some b such that $-3 < b < a$.) On the other hand, $\min[-3, 3] = -3$, even though $[-3, 3]$ is an infinite set. The rule is: we can always take the minimum and maximum of a finite ordered set, but the minimum and maximum of an infinite ordered set may or may not exist.

Definition 4.5. *(Bounds)*
*Let E be a subset of an ordered set S. If there exists an $\alpha \in S$ such that every element of E is less than or equal to α, then α is an **upper bound** of E, and E is **bounded above**.*

In symbols, E is bounded above if:

$$\exists \alpha \in S \text{ such that } \forall x \in E, \ x \leq \alpha.$$

*Similarly, if there exists a $\beta \in S$ such that every element of E is greater than or equal to β, then β is a **lower bound** of E, and E is **bounded below**.*

In symbols, E is bounded below if:

$$\exists \beta \in S \text{ such that } \forall x \in E, \ x \geq \beta.$$

 Notice that unlike with minimums and maximums, the upper or lower bound of E needn't be an element of E, it just needs to be contained in the superset S.

Example 4.6. (Bounds)
In $S = \mathbb{Q}$, the set $E = (-\infty, 3) \cap \mathbb{Q}$ has no lower bound, since for any element β of $\mathbb{Q}$, we can find an element of E that is less than β. On the other hand, E is bounded above, by any $\alpha \in \mathbb{Q}$ with $\alpha \geq 3$. Notice that $\alpha \notin E$, which doesn't matter as long as $\alpha \in S$. The infinite interval $(-\infty, \infty)$, on the other hand, is bounded neither above nor below.

Referring to the proof of Theorem 4.1, if we consider A and B to be subsets of $\mathbb{R}$, then A is bounded above by $\sqrt{2}$ (and also by anything $> \sqrt{2}$). So the set of A's upper bounds contains $\{\sqrt{2}\} \cup B$, and similarly the set of B's lower bounds contains $\{\sqrt{2}\} \cup A$.

If we ignore $\mathbb{R}$ and just consider A and B within $\mathbb{Q}$, then both A and B are bounded by the elements of each other, but not by $\sqrt{2}$ (since $\sqrt{2} \notin \mathbb{Q}$).

To further illustrate this distinction, let $E = (0, 3)$, let $S_1 = \mathbb{R}$, let $S_2 = (-3, 3)$, and let $S_3 = E$. If we consider $E \subset S_1$, then clearly E is bounded above (by any number ≥ 3) and below (by any number ≤ 0). If we consider $E \subset S_2$, then E is not bounded above, because any number greater than or equal to 3 is not in S. E is bounded below, however, by any element in $(-3, 0]$. If we consider $E \subset S_3$ (as a subset of itself), then E is bounded neither above nor below.

It is usually more precise to write "E is bounded above/below in S," instead of just "E is bounded above/below." In the case of the previous example, we would say that E is bounded above in S_1 but not in S_2 or in S_3; and E is bounded below in S_1 and in S_2 but not in S_3.

Definition 4.7. *(Supremum and Infimum)*
*Let E be a subset of an ordered set S. If there exists an upper bound α of E in S such that anything in S that is less than α is not an upper bound of E, then α is the **least upper bound** or **supremum** of E.*
  In symbols, $\alpha = \sup E$ if:

$$\alpha \in S;\ x \in E \implies x \leq \alpha;\ \text{and}\ \gamma \in S, \gamma < \alpha \implies \gamma\ \text{not an upper bound of}\ E.$$

*Similarly, if there exists a lower bound β of E in S such that anything in S that is greater than β is not a lower bound of E, then β is the **greatest lower bound** or **infimum** of E.*
  In symbols, $\beta = \inf E$ if:

$$\beta \in S;\ x \in E \implies x \geq \beta;\ \text{and}\ \gamma \in S, \gamma > \beta \implies \gamma\ \text{not a lower bound of}\ E.$$

The definition should make it apparent why the supremum is called the *least* upper bound: anything less than it is not an upper bound at all. Notice that it is called *the* supremum since it must be unique—if another one existed, one would have to be greater than the other, so one would not be a least upper bound.

As with regular upper or lower bounds, the supremum or infimum of a set E need not be an element of E, it just needs to be contained in the superset S. Thus, although a set like $(-3, 3)$ has no minimum or maximum (see Example 4.4), it does have a supremum and an infimum in $\mathbb{Q}$, namely, 3 and -3.

Example 4.8. (Suprema and Infima)
In $S = \mathbb{Q}$, the set $E = (-\infty, 3) \cap \mathbb{Q}$ has no lower bound, but one of its upper bounds, namely $3 \in S$, is in fact its supremum. To prove this, we must show that any number $\gamma < 3$ is not an upper bound of E. This is true since $\frac{\gamma+3}{2}$ (the midpoint between γ and 3) is in E, because it is a rational number which is less than 3. But $\frac{\gamma+3}{2} > \gamma$, so γ is *not* greater than or equal to every element of E.

Referring to the proof of Theorem 4.1, if we consider A and B to be subsets of $\mathbb{R}$, then A is bounded above by $\sqrt{2}$. Any real number $p < \sqrt{2}$ is not an upper bound of A (since $q = \frac{2p+2}{p+2}$ is in A and $q > p$), so $\sup A = \sqrt{2}$. Similarly, $\inf B = \sqrt{2}$.

If we ignore $\mathbb{R}$ and just consider A and B within $\mathbb{Q}$, then A does not have a supremum and B does not have an infimum, since Theorem 4.1 showed that B (the upper bounds of A) has no smallest element, and A (the lower bounds of B) has no largest element.

Try filling in the blanks in Box 4.2.

BOX 4.2

SUPREMUM AND INFIMUM OF $\{\frac{1}{n} \mid n \in \mathbb{N}\}$

In $S = \mathbb{Q}$, the set $E = \{\frac{1}{n} \mid n \in \mathbb{N}\}$ has both a supremum and an infimum.
    The supremum is $\alpha = $ _____, and we check that first α is an upper bound of E, since $\alpha \in S$, and _____ for every $x \in E$. Second, for any $\gamma < \alpha$, γ is not an _____ of E, since α is in E but $\alpha > \gamma$.
    The infimum is $\beta = $ _____, and we check that first β is a lower bound of E, since $\beta \in S$, and $x \geq \beta$ for every _____. Second, for any $\gamma > \beta$, γ is not _____ of E, since the number _____ is in E but is $< \gamma$.

Hint: To fill in that last blank, you need to find a number between β and γ that is of the form $\frac{1}{n}$ for some natural number n. Obviously $\frac{1}{n}$ is greater than β, so we just need to find a natural n such that $\frac{1}{n} < \gamma$, meaning $n > \frac{1}{\gamma}$. Well, $\frac{1}{\gamma}$ may not be a natural number, but if we round it up, it will be. We use the *ceiling* notation $\lceil x \rceil$ to mean "x rounded up to the nearest integer." Then $\lceil \frac{1}{\gamma} \rceil \in \mathbb{N}$ and $\lceil \frac{1}{\gamma} \rceil \geq \frac{1}{\gamma}$. But we actually need $n > \frac{1}{\gamma}$, not $n \geq \frac{1}{\gamma}$. We can just add 1 to make n strictly greater, so $n = 1 + \lceil \frac{1}{\gamma} \rceil$ works, giving us the number we need:

$$\frac{1}{n} = \frac{1}{1 + \lceil \frac{1}{\gamma} \rceil}.$$

Notice that the supremum in the previous example demonstrates the following: whenever an upper bound of E is an element of E, it is automatically E's supremum. This result is formalized in the following theorem.

Theorem 4.9. *(Bounds Contained in a Set)*
Let E be a subset of an ordered set S. If an upper bound of E is contained in E, then it is a least upper bound. If a lower bound of E is contained in E, then it is a greatest lower bound.

Proof. Let α be an upper bound of E, with $\alpha \in E$. For any $\gamma < \alpha$, γ cannot be an upper bound of E, since there is an element of E that is greater than γ—namely, $\alpha \in E$. Thus α is a supremum of E.
    Similarly, let β be a lower bound of E, with $\beta \in E$. For any $\gamma > \beta$, γ cannot be a lower bound of E, since there is an element of E which is less than γ—namely, $\beta \in E$. Thus β is an infimum of E. □

Of course, when a subset does not contain any of its upper bounds, it may or may not have a supremum.
    Get excited! We're about to define one of the most important concepts in real analysis.

Definition 4.10. *An ordered set S has the **least upper bound property** if, for every nonempty subset E which is bounded above in S, sup E exists in S.*

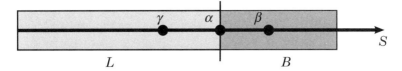

Figure 4.1. The subsets B and L, represented as intervals on the line S.

Example 4.11. Does $\mathbb{Q}$ have the least upper bound property? Referring to the proof of Theorem 4.1, Example 4.8 demonstrated that A is a nonempty subset of $\mathbb{Q}$ which is bounded above in $\mathbb{Q}$ (by all the elements of B). But A has no supremum in $\mathbb{Q}$ (since $\sqrt{2} \notin \mathbb{Q}$). Thus $\mathbb{Q}$ does *not* have the least upper bound property. Oh, well. (Actually, the fact that $\mathbb{Q}$ is missing the least upper bound property is the main motivation for defining $\mathbb{R}$.)

"If there's a least upper bound property, then shouldn't there also be a greatest lower bound property?", you might ask? Well, yes—but as it turns out, the two are equivalent!

Theorem 4.12. *(Supremum of Lower Bounds)*
Let S be an ordered set with the least upper bound property. Then S also has the greatest lower bound property—namely, for every nonempty subset B which is bounded below in S, $\inf B$ exists in S.

Furthermore, if we denote L as the set of all lower bounds of B, then $\inf B = \sup L$.

This theorem actually goes beyond proving that one property implies the other—it tells us that in a set with the least upper bound property, the infimum of any subset is just the supremum of that subset's lower bounds.

Proof. Take the set L of all lower bounds of B. We will start by showing that the number $\sup L$ exists in S and then show that actually $\sup L = \inf B$.

To show $\sup L$ exists in S, we want to take advantage of the least upper bound property, so we want to show that L is a nonempty subset which is bounded above in S. The fact that B is bounded below in S tells us L has at least one element in S. How do we show that L is bounded above in S? Every element of L is less than or equal to every element of B, so every element of B is an upper bound of every element of L. Since B is a nonempty subset of S, L has at least one upper bound in S. Now we apply the least upper bound property to see that $\sup L$ exists in S; let's call it α.

For the next part of the proof, refer to Figure 4.1.

For α to equal $\inf B$, it must first be a lower bound of B. Because α is the supremum of L, if any number γ is less than α, then γ is not an upper bound of L—meaning γ must be less than some element of L. Thus γ is smaller than a lower bound of B, so γ cannot be in B. We have shown that any number less than α cannot be in B, so α must be $\leq x$ for every $x \in B$, so α is a lower bound of B.

Now α is a lower bound of B, but if any number β is greater than α, β cannot belong to L, since α is an upper bound of L. Meaning, any $\beta > \alpha$ is not a lower bound of B, so indeed $\alpha = \inf B$. □

The converse of the previous theorem completes the proof that the least upper bound property and the greatest lower bound property are equivalent.

34 • Chapter 4

Theorem 4.13. *(Infimum of Upper Bounds)*
Let S be an ordered set with the greatest lower bound property. Then S also has the least upper bound property—namely, for every nonempty subset B which is bounded above in S, sup B exists in S.

Furthermore, if we denote U as the set of all upper bounds of B, then $\sup B = \inf U$.

Proof. This proof is just a mirror of the previous one. After filling in the blanks in Box 4.3, draw a picture similar to Figure 4.1 before finishing the second paragraph.

BOX 4.3

> PROVING THEOREM 4.13
>
> To show $\inf U$ exists in S, we want to take advantage of the _____ property. The fact that B is bounded above in S tells us U has at least one element in S, so U is _____. Every element of U _____ every element of B, so every element of B is a _____ of every element of U. B is a nonempty subset of S, so U has at least one lower bound in S. Now we apply the greatest lower bound property to see that $\inf U$ exists in S; let's call it α.
>
> If $\gamma > \alpha$, then γ is not a _____ of U, meaning γ must be greater than some element of U, so $\gamma \notin B$. Thus α must be _____ x for every $x \in B$, so α is an upper bound of B. If any number β is less than α, then β _____ U, since α is an upper bound of U. Meaning, any $\beta < \alpha$ is not an upper bound of B, so indeed $\alpha = $ _____.

□

Suprema and infima are constantly called on in real analysis. A bounded infinite set may not have a minimum or maximum, but it will always have a supremum or infimum, which is usually our best way of describing the narrowest possible bounds. Suprema and infima will keep coming back, so it's worth making the effort now to really understand their definitions and how to manipulate them in proofs. (If you hate them and want them to go away, trying saying "soupy" whenever you read sup E; it might help you feel better.)

We want $\mathbb{R}$ to be an ordered superset of $\mathbb{Q}$ with the least upper bound/greatest lower bound property. We also want it to work with addition, multiplication, and more, which are properties of a *field*—we'll define what that means in the next chapter.

CHAPTER 5

The Real Field

After learning the definition of an ordered field, we can finally understand what the real numbers are. As a field, $\mathbb{R}$ has three important characteristics: the Archimedean property, the density of $\mathbb{Q}$, and the existence of roots. We will prove each of these, and will frequently take advantage of them henceforth.

Definition 5.1. *(Field)*
*Any set F with two operations, addition and multiplication, is a **field** if it satisfies the following **field axioms**:*

- A1. *(Closed under Addition) If $x, y \in F$, then $x + y \in F$.*
- A2. *(Commutative under Addition) $x + y = y + x$, $\forall x, y \in F$.*
- A3. *(Associative under Addition) $(x + y) + z = x + (y + z)$, $\forall x, y, z \in F$.*
- A4. *(Additive Identity) F contains an element 0 such that $x + 0 = x$, $\forall x \in F$.*
- A5. *(Additive Inverse) For every $x \in F$ there exists an element $-x \in F$ such that $x + (-x) = 0$.*

- M1. *(Closed under Multiplication) If $x, y \in F$, then $xy \in F$.*
- M2. *(Commutative under Multiplication) $xy = yx$, $\forall x, y \in F$.*
- M3. *(Associative under Multiplication) $(xy)z = x(yz)$, $\forall x, y, z \in F$.*
- M4. *(Multiplicative Identity) F contains an element 1 (with $1 \neq 0$) such that $1x = x$, $\forall x \in F$.*
- M5. *(Multiplicative Inverse) For every $x \in F$ with $x \neq 0$, there exists an element $\frac{1}{x} \in F$ such that $(x)\frac{1}{x} = 1$.*

- D. *(Distributive Law) $x(y + z) = xy + xz$, $\forall x, y, z \in F$.*

If we take any set together with our conventional definitions of addition and multiplication, most of these axioms follow naturally. For example, we know that the multiplicative identity is the number 1, since 1 times anything is itself. If we want, we could define some new, different versions of addition and multiplication on a set and try to get them to satisfy the field axioms, but we don't really want to do that. Studying structures such as fields in an abstract manner is the focus of *algebra*, an advanced area of mathematics which has nothing to do with "if $3x + 2 = 8$, find x."

For our purposes, since every field we will consider contains ordinary numbers with normal addition and multiplication, we can mostly take for granted the axioms *A2–A5*, *M2–M5*, and *D*. However, the axioms of closure, *A1* and *D1*, are not trivial in many cases and must be proved explicitly if we want to show that something is a field.

Example 5.2. (Fields)
The rational numbers, along with normal addition and multiplication, constitute a field. Let's check that the axioms of closure hold. For any $x, y \in \mathbb{Q}$, we can write $x = \frac{a}{b}$ and $y = \frac{c}{d}$ for some $a, b, c, d \in \mathbb{Z}$. Then

$$x + y = \frac{a}{b} + \frac{c}{d} = \frac{ad + bc}{bd},$$

which is also a rational number, and

$$xy = \frac{ac}{bd},$$

which is also a rational number.

On the other hand, $\mathbb{N}$ is *not* a field—no element has an additive inverse, since $\mathbb{N}$ does not contain negative integers. $\mathbb{Z}$ is also not a field—no element has a multiplicative inverse, since $\mathbb{Z}$ does not contain fractions.

The set $S = \{0, 1, 2, 3, 4\}$, along with normal addition and multiplication, is not a field. The numbers 2 and 3 are in S, but $2 + 3 = 5$ is not in S, so S is not closed under addition. Similarly, $(2)(3) = 6 \notin S$, so S is not closed under multiplication either.

But let's say we take $T = S$ modulo 5, meaning S where we set the number 5 equal to 0. Then $10 = 5 + 5 = 0 + 0$, and likewise any multiple of 5 also equals 0. In this case, $2 + 3 = 5 = 0$ is in T. So is any sum or product x of elements of T, since if $x \geq 5$, we can write it as $x = 5n + m = m$ for some $m, n \in \mathbb{N}$ with $m < 5$. This strange field T is usually denoted $\mathbb{Z}_5$.

Using only the field axioms, we can prove many basic properties of all fields, such as $-(-x) = x$, $0x = 0$, $xy = 1 \implies y = \frac{1}{x}$ if $(x \neq 0)$, and so on. The proofs of these properties just use the field axioms multiple times to rearrange terms.

Definition 5.3. *(Ordered Field)*
An **ordered field** is a field F that is also an ordered set, satisfying the following axioms:

O1. If $y < z$, then $x + y < x + z$, $\forall x, y, z \in F$.
O2. If $x > 0$ and $y > 0$, then $xy > 0$, $\forall x, y \in F$.

Go back to Definition 4.2 to review the meaning of an ordered set.

The axioms for ordered fields are pretty intuitive, and we could use them to prove some basic properties, such as $x > 0, y < z \implies xy < xz$, $0 < x < y \implies 0 < \frac{1}{y} < \frac{1}{x}$, and so on. (We'll skip over these simple proofs, because time is money, and money can buy you a blank notebook to try writing these proofs yourself.)

Now we are ready to define the real numbers!

Definition 5.4. *(Real Numbers)*
The set of **real numbers** $\mathbb{R}$ is an ordered field which has the least upper bound property and contains $\mathbb{Q}$.

In other words, ℝ satisfies all the axioms of an ordered field, and fills in all the "holes" in the rational numbers.

Just because we can define ℝ, though, doesn't mean it exists. Different methods of teaching real analysis solve this problem in different ways. Some textbooks just assume that this ℝ exists, and call this assumption "the axiom of completeness"—here, *completeness* is just another word for the least upper bound/greatest lower bound property. It is possible, though, to *prove* that this ℝ exists (after assuming that ℚ exists). One (fairly laborious) proof, uses something called *Dedekind cuts*, which you can look up if you're interested.

In addition to being complete and containing ℚ, the real numbers have some very useful properties, which are explored in the following three theorems.

Theorem 5.5. *(Archimedean Property of ℝ)*
Given any positive real number, we can find a natural number such that the product of the two is as large as we want.
In symbols:

$$\forall x, y \in \mathbb{R} \text{ with } x > 0, \exists n \in \mathbb{N} \text{ such that } nx > y.$$

In other words, the Archimedean property asserts that you can always find a natural number greater than any ratio of real numbers. A commonly used corollary is $\forall y \in \mathbb{R}$, $\exists n \in \mathbb{N}$ such that $n > y$ (which we get by setting $x = 1$ in the original theorem).

Of course, ℚ also has the Archimedean property. For any $x, y \in \mathbb{Q}$ with $x > 0$, we can write $x = \frac{a}{b}$ and $y = \frac{c}{d}$ where $a, b, c, d \in \mathbb{N}$. Let $n = 2bc$, so clearly $n \in \mathbb{N}$, and we have

$$nx = (2bc)\frac{a}{b} = 2ac = (2ad)\frac{c}{d} = 2ady > y.$$

(Here we assumed $y > 0$. If $y \leq 0$, just let $n = 1$, so that $nx = x > y$ since $x > 0$.)

Proving the Archimedean property for the real numbers is a little trickier, since most real numbers don't have a simple representation like fractions do. Instead, we'll take advantage of what makes ℝ so special: the least upper bound property.

Proof. Let's take this opportunity to demonstrate the two-step proof process. First, we'll break the proof down step-by-step by using the definitions to figure out what the crux of the problem is. Second, we'll take what we have and write it up in a nice, linear fashion.

Step 1. Let A be the set of each possible nx, for every natural number n. The Archimedean property asserts that some element of A is greater than y. If we want to apply the least upper bound property, the set A can't really help us, since it is not bounded above—unless we assume the Archimedean property is false. Then no element of A is greater than y, meaning y is an upper bound of A. This seems like a plausible direction to go in, so let's do a proof by contradiction.

Assuming the Archimedean property is false, $A = \{nx \mid n \in \mathbb{N}\}$ is bounded above by y. Now A is a nonempty subset of ℝ, so by Definition 4.10, sup A exists in ℝ; write $\alpha = \sup A$ for convenience. Now which available facts have we not used yet? Basically, all we have left to apply is the definition of a supremum: if $\gamma < \alpha$, then γ is not an upper bound of A. That means γ is less than some element of A, so there exists some natural number m such that $\gamma < mx$.

Does that help? Sort of. It looks like a contradiction we might want to end up with is "α is not an upper bound of A." So if we show that some natural number times x is greater than α, we'll be done (since this contradicts the fact that $\alpha = \sup A$). We have $\gamma < mx$, so if we can put γ in terms of α, that inequality will be more useful. The restriction on γ is that it must be less than α, so let's write $\gamma = \alpha - k$ for some $k > 0$. Then $\gamma < mx \implies \alpha < mx + k$.

Now the goal is to put $mx + k$ into the form of nx for some natural n. If $k = cx$ for any $c \in \mathbb{N}$, then $mx + k = mx + cx = (m + c)x$, which is indeed a natural number times x. Actually the simple case $c = 1$ works, so $k = x$ is exactly what we need. Now $\alpha < (m + 1)x$, so α is less than some element of A, which contradicts the fact that $\alpha = \sup A$. Thus the Archimedean property for the real numbers must be true.

Step 2. After all that thinking, it didn't take very many steps to prove the theorem. We can now write it up formally.

Assume that $\mathbb{R}$ does not have the Archimedean property. Then $A = \{nx \mid n \in \mathbb{N}\}$ is bounded above by y, so because $\mathbb{R}$ has the least upper bound property, $\alpha = \sup A$ exists in $\mathbb{R}$. Since $x > 0$, we have $\alpha - x < \alpha$, so $\alpha - x$ cannot be an upper bound of A. Then $\exists m \in \mathbb{N}$ such that $\alpha - x < mx$, so $\alpha < (m + 1)x$. However $(m + 1)x \in A$, so α is not an upper bound of A, which is a contradiction.

If we want, we could shorten the proof even more by putting almost everything in symbols:

$$\neg(\exists n \text{ such that } nx > y) \implies A = \{nx \mid n \in \mathbb{N}\} \leq y$$
$$\implies \alpha = \sup A \in \mathbb{R}$$
$$\implies \exists m \in \mathbb{N} \text{ such that } \alpha - x < mx$$
$$\implies \alpha < (m + 1)x$$
$$\implies \bot.$$

(In logic, the symbol $\bot$ means "contradiction.")

□

Theorem 5.6. *($\mathbb{Q}$ Is Dense in $\mathbb{R}$)*
Between every two real numbers, there is at least one rational number.
In symbols:

$$x, y \in \mathbb{R} \text{ with } x < y \implies \exists p \in \mathbb{Q} \text{ such that } x < p < y.$$

This property of $\mathbb{Q}$ in $\mathbb{R}$ is known as *density*; we say that $\mathbb{Q}$ is *dense* in $\mathbb{R}$.

We already know that $\mathbb{R}$ is dense in itself (we can always find a real number between any two given real numbers). Why? For any $x, y \in \mathbb{R}$ with $x < y$, let $z = \frac{x+y}{2}$ (the midpoint between x and y), so that $x < z < y$.

By this same logic, we see that $\mathbb{Q}$ is dense in itself, since the midpoint $p = \frac{q+r}{2}$ is always between q and r in $\mathbb{Q}$. Similarly, $\mathbb{Q}$ is also dense in $\mathbb{N}$ (where $p = \frac{m+n}{2}$ is the rational midpoint between two natural numbers m and n).

On the other hand, $\mathbb{N}$ is not dense in itself, since there is no natural number between 2 and 3. Similarly, $\mathbb{N}$ is not dense in $\mathbb{Q}$, because there is no natural number between $\frac{1}{3}$ and $\frac{1}{2}$.

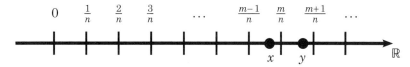

Figure 5.1. We must choose n large enough so that the increments $\frac{1}{n}$ do not skip over the interval (x, y). We must choose m to be the smallest integer greater than nx.

 Note that density guarantees not only that there is *one* p between x and y but that there are *infinitely* many rational numbers between x and y. Once we have $x < p < y$, we can just apply the property again to find a $q \in \mathbb{Q}$ such that $x < q < p$ and repeat this step an infinite number of times.

One particular consequence of this theorem is that any open interval (a, b) or closed interval $[a, b]$ contains an infinite number of points, as do $(a, b) \cap \mathbb{Q}$ and $[a, b] \cap \mathbb{Q}$.

 Proof. We want to find a rational number p between x and y—meaning we want to find some $m, n \in \mathbb{Z}$ such that $x < \frac{m}{n} < y$. By applying the Archimedean property from Theorem 5.5 a bunch of times, we hope to find the perfect m and n, as shown in Figure 5.1.

Part 1. First, we need to make sure $n \in \mathbb{N}$ is large enough so that at least one of the fractions $\frac{1}{n}, \frac{2}{n}, \frac{3}{n}, \ldots$ falls between x and y. If n were too small, then the increment between the fractions $\frac{1}{n}, \frac{2}{n}, \frac{3}{n}, \ldots$ might be too large, making them skip over the interval (x, y).

In other words, we need $\frac{1}{n}$ to be less than $y - x$. If we rearrange the inequality to look like $n(y - x) > 1$, we can apply the Archimedean property to find such an $n \in \mathbb{N}$. This gives us $ny > 1 + nx$, which we will use later on.

Part 2. Second, we need to make sure $m \in \mathbb{Z}$ is the smallest integer greater than nx, so that $nx < m < ny$.

Since we want to "trap" nx between two consecutive integers, we need to show that $\exists m \in \mathbb{Z}$ such that $m - 1 \leq nx < m$. There are three possible cases.

> *Case 1.* If $nx > 0$, then we can apply the Archimedean property to show that there exists at least one $m_1 \in \mathbb{N}$ such that $m_1 > nx$. (Here, the variable names get a little confusing. We are applying Theorem 5.5, where $x \in \mathbb{R}$ is 1, $y \in \mathbb{R}$ is nx, and $n \in \mathbb{N}$ is m_1.) Thus, the set $\{m_1 \in \mathbb{N} \mid m_1 > nx\}$ is nonempty.
>
> There is an axiom from number theory called *the well-ordering principle*, which states that every nonempty subset of $\mathbb{N}$ has a smallest element. (This should make intuitive sense, because $\mathbb{N}$ is bounded below by the number 1.) The well-ordering principle tells us that $\{m_1 \in \mathbb{N} \mid m_1 > nx\}$ has a smallest element m. Since m is the smallest integer greater than nx, $m - 1$ must be $\leq nx$. Thus $m - 1 \leq nx < m$.
>
> *Case 2.* If $nx = 0$, then $0 \leq nx < 1$. So with $m = 1$, we have $m - 1 \leq nx < m$.
>
> *Case 3.* If $nx < 0$, then we apply Case 1 on $-nx$ to yield a natural number m_2 such that $m_2 - 1 \leq -nx < m_2$. Therefore $m_2 < nx \leq 1 - m_2$. If $nx = 1 - m_2$, then $m = 2 - m_2$ satisfies $m - 1 \leq nx < m$. Otherwise, $nx < 1 - m_2$, so $m = 1 - m_2$ satisfies $m - 1 \leq nx < m$.

Now we have an integer m such that $nx < m$ and $m \leq 1 + nx$. If we combine this with the inequality from Part 1, we get

$$nx < m \leq 1 + nx < ny.$$

Since $n > 0$, we can divide by n to get $x < \frac{m}{n} < y$, which is exactly what we needed. □

Corollary 5.7. *(The Irrational Numbers Are Dense in $\mathbb{R}$)*
Between every two real numbers, there is at least one irrational number.
   In symbols:

$$x, y \in \mathbb{R} \text{ with } x < y \implies \exists p \notin \mathbb{Q} \text{ such that } x < p < y.$$

Proof. Let $a = \frac{x}{\sqrt{2}}$ and $b = \frac{y}{\sqrt{2}}$. By Theorem 5.6, there is a rational number q such that $a < q < b$. Then $x < \sqrt{2}q < y$, so $p = \sqrt{2}q$ is exactly what we needed. Note that p is indeed irrational, since if it were rational then $\sqrt{2} = \frac{p}{q}$ would be rational. □

The following theorem (the last of our terrific trio) deals with roots. You know what these are: usually we write $y = \sqrt[n]{x}$ or $y = x^{\frac{1}{n}}$.

Theorem 5.8. *(Existence of Roots in $\mathbb{R}$)*
Every positive real number has a unique positive nth root, for any $n \in \mathbb{N}$.
   In symbols:

$$\forall x \in \mathbb{R} \text{ with } x > 0, \forall n \in \mathbb{N}, \exists y \in \mathbb{R} \text{ unique, such that } y > 0 \text{ and } y^n = x.$$

Note that even-numbered roots ($\sqrt{x}$, $\sqrt[4]{x}$, $\sqrt[6]{x}$, etc.) signify two numbers in $\mathbb{R}$, namely, $+y$ and $-y$. The theorem asserts that there is one and only one *positive* real root.

Proof. The uniqueness of y is the easiest part of the proof, so we'll start there. For any two different positive real numbers, the fact that they are different means one must be greater than the other. If there were two positive real roots y_1 and y_2 such that $y_1^n = x$ and $y_2^n = x$, we would have $0 < y_1 < y_2$. But then $0 < y_1^n < y_2^n$, meaning $0 < x < x$, which is impossible. Thus only one positive real root can exist.

To prove that $\sqrt[n]{x}$ exists in $\mathbb{R}$, let's first figure out our game plan and then write it up formally.

Step 1. We'll use the least upper bound property of $\mathbb{R}$. Remember the discussion of Theorem 4.1 in Example 4.8? We ended up seeing that $\sqrt{2}$ is the supremum of the set A and the infimum of the set B. If we construct the more general set E corresponding to the numbers less than $\sqrt[n]{x}$, we can apply the least upper bound property to show that $y = \sup E$ exists in $\mathbb{R}$. Since we don't know that $\sqrt[n]{x}$ exists yet, we can't use it when defining E; so we define E this way:

$$E = \{t \in \mathbb{R} \mid t > 0 \text{ and } t^n < x\}.$$

The crux of this problem, then, is to show that indeed $y = \sqrt[n]{x}$. In Theorem 4.1, we found the "magic number" $q = \frac{2p+2}{p+2}$ to demonstrate that any $p < \sqrt{2}$ cannot be an upper bound of A (since $q \in A$ but $q > p$). But for generic roots of positive real numbers, how do we show that any number less than $\sqrt[n]{x}$ is not an upper bound of E?

Our strategy will be to show that $y = \sup E$ cannot be less than $\sqrt[n]{x}$ and it cannot be greater than $\sqrt[n]{x}$, so it must equal $\sqrt[n]{x}$. If $y^n < x$, we should end up with the statement "y is not an upper bound of E," which contradicts the fact that $y = \sup E$. If $y^n > x$, we should end up with the statement "some number less than y is an upper bound of E," which contradicts the fact that $y = \sup E$.

First we need to show that E satisfies the requirements of the least upper bound property. In other words, we need E to be nonempty and bounded above. To show that there is at least one element in E, we want to find a $t > 0$ such that $t^n < x$. We cannot simply pick $t = \frac{\sqrt[n]{x}}{2}$, since we haven't yet proved that $\sqrt[n]{x}$ is a real number. Instead, if we can find a t that is less than x and with $t^n \leq t$, we'll be set (since then $t^n \leq t < x$). Without too much effort, we can see that anything like $t = \frac{x}{x+1}$ meets those requirements.

To show that E has an upper bound, we want to find a u such that $t > u \implies t^n \geq x$. Again, we cannot simply pick $u = 2\sqrt[n]{x}$, so we'll look for something with $u > x$ and $u^n \geq u$ (so that $t > n \implies t^n > u^n \geq u > x$). We can see that anything like $u = x + 1$ does the trick (remember $x > 0$).

Since $\mathbb{R}$ has the least upper bound property, $y = \sup E$ exists.

Now let's see what happens when $y^n < x$. To get our contradiction "y is not an upper bound of E," we want to find an element of E that is greater than y, meaning some t such that $t^n < x$ but $t > y$. Letting $t = y + h$, we need to find a real $h > 0$ with $(y + h)^n < x$.

When faced with a tough situation like this, we can appeal to the information we have about powers in general. Using some good old-fashioned algebraic manipulation, we can calculate

$$(b-a)\sum_{k=1}^{n} b^{n-k}a^{k-1} = (b-a)(b^{n-1} + b^{n-2}a + \ldots + ba^{n-2} + a^{n-1})$$

$$= (b^n + b^{n-1}a + \ldots + b^2 a^{n-2} + ba^{n-1})$$
$$\quad - (b^{n-1}a + b^{n-2}a^2 + \ldots + ba^{n-1} + a^n)$$

$$= b^n - a^n.$$

Furthermore, when $0 < a < b$, we have

$$\sum_{k=1}^{n} b^{n-k}a^{k-1} = b^{n-1}\sum_{k=1}^{n} \left(\frac{a}{b}\right)^{k-1}$$

$$< b^{n-1}\sum_{k=1}^{n}(1)^{k-1}$$

$$= nb^{n-1},$$

which gives us the inequality

$$b^n - a^n < (b-a)nb^{n-1}.$$

If we substitute $b = y + h$ and $a = y$ (so indeed $0 < a < b$), we have

$$(y + h)^n - y^n < hn(y + h)^{n-1}.$$

Because we are trying to show $(y + h)^n < x$, it will now be sufficient to show $hn(y + h)^{n-1} < x - y^n$ (you can chain the inequalities to confirm this).

🚩 *Wait.* Does pulling that inequality out of thin air seem like a cheap shot? Perhaps. In complex proofs like these, I don't think it's possible to explain how you would come up with each and every step on your own. My guess is that the mathematicians who figured out this proof spent a long time playing around with power expansions and had to try out many inequalities before chancing on the one that worked. Most of the time, pure math is about exploring and trying new things until you manage to find the silver bullet. On the other hand, I don't think you would be expected to pull off something like this on an assignment or an exam (at least not without an explicit hint).

Back to the proof. Remember we are trying to find a real $h > 0$ such that $hn(y + h)^{n-1} < x - y^n$. Our job is simplified if we choose $h < 1$, since then $hn(y + h)^{n-1} < hn(y + 1)^{n-1}$. Now the inequality is satisfied as long as

$$h < \frac{x - y^n}{n(y + 1)^{n-1}}.$$

We still haven't made use of the assumption $y^n < x$, and here is where it finally comes into play: the assumption $y^n < x$ guarantees that the fraction is positive, so that $\frac{x-y^n}{n(y+1)^{n-1}}$ is a valid positive real number.

Thus for any real h with

$$0 < h < \min\left\{1, \frac{x - y^n}{n(y + 1)^{n-1}}\right\},$$

we have $(y + h)^n < x$. Then $y + h \in E$ even though y is an upper bound, which is a contradiction.

Next, let's see what happens when $y^n > x$. To get our contradiction "some number less than y is an upper bound of E," we want to find a real $k > 0$ such that $y - k$ is an upper bound of E. Note that k must also be less than y, since $y - k$ should be positive. So we want to prove that any number greater than $y - k$ is not in E, meaning $t > y - k \implies t^n \geq x$.

If we take $t > y - k$, we get

$$y^n - t^n < y^n - (y - k)^n,$$

whose right-hand side we want to show is $\leq y^n - x$. Then we will have $y^n - t^n \leq y^n - x$, so $x \leq t^n$. As it turns out, we can use the same inequality as in the previous part! Plugging $b = y$ and $a = y - k$ into $b^n - a^n < (b - a)nb^{n-1}$, we have

$$y^n - (y - k)^n < kny^{n-1}.$$

(since $0 < a < b$, as required). Then

$$k = \frac{y^n - x}{ny^{n-1}}$$

is exactly what we need, as long as we check that $0 < k < y$. We use the critical assumption $y^n > x$ to see that k must be positive, and indeed k also satisfies

$$k = \frac{y^n - x}{ny^{n-1}} < \frac{y^n}{ny^{n-1}} = \frac{y}{n} \le y.$$

Thus we see that any number greater than $y - k$ is not in E, so $y - k$ is an upper bound even though y is the supremum, which is a contradiction.

Step 2. Whew! That was a lot of figuring out, but the actual proof shouldn't be too long. As you read the following summary, refer to the text from Step 1 to make sure you understand how each assertion follows from the previous one.

Let

$$E = \{t \in \mathbb{R} \mid t > 0 \text{ and } t^n < x\}.$$

Then E is nonempty, since

$$t = \frac{x}{x+1} \implies t^n \le t \text{ (since } x < x+1, \text{ so } t < 1)$$
$$\text{and } t < x \text{ (since } x < x + x^2)$$
$$\implies 0 < t^n \le t < x$$
$$\implies t \in E.$$

E is bounded above, since with $u = x + 1$,

$$t > u \implies t^n \ge t \text{ (since } x + 1 > x, \text{ so } t > 1)$$
$$\implies t^n \ge t > u > x$$
$$\implies t \notin E$$
$$\implies u \text{ is an upper bound.}$$

So $y = \sup E$ exists in $\mathbb{R}$ by the least upper bound property.

Note that for any $0 < a < b$, we have

$$b^n - a^n = (b - a) \sum_{k=1}^{n} b^{n-k} a^{k-1} < (b - a) n b^{n-1}.$$

Assume $y^n < x$, and choose

$$h < \min \left\{ 1, \frac{x - y^n}{n(y+1)^{n-1}} \right\}.$$

44 • Chapter 5

Then there exists such an h that is also positive, so that $0 < y < y + h$, and thus

$$(y+h)^n - y^n < hn(y+h)^{n-1}$$
$$< hn(y+1)^{n-1}$$
$$< x - y^n.$$

Now $(y+h)^n < x$, so $y + h \in E$ even though y is an upper bound of E. We have a contradiction, so it must be the case that $y^n \geq x$.

Assume $y^n > x$, and choose

$$k = \frac{y^n - x}{ny^{n-1}}.$$

Then $0 < k < y$ (as shown above), so for any $t > y - k$, we have

$$y^n - t^n < y^n - (y-k)^n$$
$$< kny^{n-1}$$
$$= y^n - x.$$

Now $x < t^n$, so $t \notin E$, so $y - k$ is an upper bound of E even though y is the supremum of E. We have a contradiction, so it must be the case that $y^n \leq x$.

Therefore y^n must equal x, meaning $\sqrt[n]{x} \in \mathbb{R}$. $\square$

Corollary 5.9. *(The Root Operation Is Distributive)*
Taking roots of positive real numbers distributes across multiplication.
In symbols:

$$\forall a, b \in \mathbb{R} \text{ with } a, b > 0, \forall n \in \mathbb{N}, (ab)^{\frac{1}{n}} = a^{\frac{1}{n}} b^{\frac{1}{n}}.$$

Why is this corollary not completely obvious? The field axiom $M2$ asserts that $ab = ba$, which implies

$$(ab)^n = (ab)(ab)\ldots(ab) = (aa\ldots a)(bb\ldots b) = a^n b^n.$$

Keep in mind, though, that raising a number x to the nth power is completely different from taking its nth root. The former is just shorthand for doing multiplication; the latter represents finding a positive real y such that $y^n = x$ (which, until just now, we didn't know was possible).

Proof. By Theorem 5.8, $\sqrt[n]{a}$ and $\sqrt[n]{b}$ exist, so we can write $a = (\sqrt[n]{a})^n$ and $b = (\sqrt[n]{b})^n$. Then

$$ab = (\sqrt[n]{a})^n (\sqrt[n]{b})^n = (\sqrt[n]{a}\sqrt[n]{b})^n,$$

by field axiom $M2$.

Theorem 5.8 also asserts uniqueness. There can only be one positive real y such that $y^n = ab$, so the fact that $(\sqrt[n]{a}\sqrt[n]{b})^n = ab$ implies $\sqrt[n]{a}\sqrt[n]{b} = \sqrt[n]{ab}$. We can also write this equality as $(ab)^{\frac{1}{n}} = a^{\frac{1}{n}} b^{\frac{1}{n}}$. $\square$

The Real Field • 45

We'll finish our characterization of the field of real numbers with a clarification of how to work with infinity. The elements $+\infty$ and $-\infty$ are *not* real numbers, but we can introduce them as symbols to use.

Definition 5.10. *(Extended Real Number System)*
The **extended real number system** is $\mathbb{R} \cup \{+\infty, -\infty\}$. It has the same ordering as $\mathbb{R}$, with the additional rule $-\infty < x < +\infty$ for every $x \in \mathbb{R}$.

The symbols $+\infty$ and $-\infty$ follow these conventions, for every $x \in \mathbb{R}$:

$$x + \infty = +\infty, \ x - \infty = -\infty;$$

$$\frac{x}{+\infty} = \frac{x}{-\infty} = 0;$$

$$x > 0 \implies x(+\infty) = +\infty, \ x(-\infty) = -\infty;$$

$$x < 0 \implies x(+\infty) = -\infty, \ x(-\infty) = +\infty.$$

Note that no conventions are given for computing

$$\infty - \infty, \ \frac{\infty}{\infty}, \ \frac{0}{\infty}, \ \frac{\infty}{0}, \text{ or } (\infty)(\infty);$$

these are all undefined quantities.

The extended real number system satisfies the ordered field axioms of Definition 5.3, but it does *not* satisfy the field axioms of Definition 5.1 (since, for example, $+\infty$ does not have a multiplicative inverse). Again, the extended reals is not a field, but it will nonetheless come in handy once in a while.

One interesting property is that every subset of the extended reals is bounded. If a subset $E \subset \mathbb{R}$ is not bounded above in $\mathbb{R}$, it *is* bounded above in $\mathbb{R} \cup \{+\infty, -\infty\}$—that is, it is bounded above by the element $+\infty$. Then, because $\mathbb{R}$ has the least upper bound property, E must have a supremum, and this supremum is $+\infty$. Similarly, if $E \subset \mathbb{R}$ is not bounded below in $\mathbb{R}$, its lower bound and its infimum in the extended reals is $-\infty$.

This was an intense chapter! But it feels good to have a solid understanding of what the real field is.

Coming up next, we'll generalize $\mathbb{R}$ to the vector space $\mathbb{R}^k$ and explore its properties. If you've always wondered what an "imaginary number" is, turn the page.... Or just use your imagination.

CHAPTER 6

Complex Numbers and Euclidean Spaces

Why would we bother with complex numbers in real analysis? Don't they involve "imaginary" numbers, which are clearly not "real" numbers? Isn't there a whole separate field of study called "complex analysis"? That's all true. But it's also true that complex numbers are just 2-dimensional real numbers (with some special operations tacked on).

We'll start by defining complex numbers as real 2-vectors like (a, b) to prove that they make up a field. Then we can show how these complex numbers are actually the same as the imaginary numbers of the form $a + bi$, which you may have seen before. After proving some properties of complex numbers, we'll be able to discover similar properties for real vectors of any size, which make up sets called *Euclidean spaces*.

Definition 6.1. *(k-Vector)*
A ***k-vector*** or ***k-dimensional vector*** *is an ordered set of numbers denoted by* $(x_1, x_2, \ldots, x_k)$. *The "order" matters since* $(x_1, x_2) \neq (x_2, x_1)$ *unless* $x_1 = x_2$.

For example, a 2-dimensional vector of real numbers would be (a, b), where $a, b \in \mathbb{R}$. Two 2-vectors $x = (a, b)$ and $y = (c, d)$ are equal if and only if $a = c$ and $b = d$.

Definition 6.2. *(Complex Number)*
A ***complex number*** *is a 2-vector of real numbers, with the operations addition and multiplication as defined by:*

$$x + y = (a + c, b + d),$$

$$xy = (ac - bd, ad + bc).$$

The set of all complex numbers is denoted by $\mathbb{C}$. The set of all complex numbers *without* the complex addition and complex multiplication operations is just the set of all 2-dimensional real numbers, denoted by $\mathbb{R}^2$.

Notice that like rational numbers, complex numbers are ordered pairs. But $(a, b) = (c, d)$ if and only if $a = c$ and $b = d$, unlike rational numbers where $\frac{a}{b}$ and $\frac{c}{d}$ can be equal if, for example, $a = 2c$ and $b = 2d$.

 Distinction. The complex number $(-3, 3)$ is not the same as the open interval $(-3, 3)$. The former is a pair of two real numbers (-3 and 3), whereas the latter is the set of all real numbers between -3 and 3. This ambiguity seems needlessly confusing, but it doesn't usually pose a problem; you should be able to tell a complex number from an open interval given the context.

Theorem 6.3. *(ℂ Is a Field)*
The set of all complex numbers is a field.
   Specifically, $(0, 0)$ is the additive identity and $(1, 0)$ is the multiplicative identity. For any complex number $x = (a, b)$, $-x = (-a, -b)$ is its additive inverse; and if $x \neq (0, 0)$ then $\frac{1}{x} = \left(\frac{a}{a^2+b^2}, \frac{-b}{a^2+b^2}\right)$ is its multiplicative inverse.

Proof. We want to show that each of the field axioms is true for ℂ, using the identities and inverses defined in the theorem. Note that in the proof of each axiom, we make use of the fact that ℝ satisfies that same axiom. Let $x = (a, b)$, $y = (c, d)$, and $z = (e, f)$ be complex numbers.

   A1. (Closed under Addition) $x + y = (a + c, b + d)$. Since ℝ is closed under addition, $a + c \in \mathbb{R}$ and $b + d \in \mathbb{R}$, so indeed
   $$(a + c, b + d) \in \mathbb{C}.$$

   A2. (Commutative under Addition)
   $$x + y = (a + c, b + d)$$
   $$= (c + a, d + b) = y + x.$$

   A3. (Associative under Addition)
   $$(x + y) + z = (a + c, b + d) + (e, f)$$
   $$= ((a + c) + e, (b + d) + f)$$
   $$= (a + (c + e), b + (d + f))$$
   $$= (a, b) + (c + e, d + f) = x + (y + z).$$

   A4. (Additive Identity)
   $$x + 0 = (a, b) + (0, 0)$$
   $$= (a, b) = x.$$

   A5. (Additive Inverse)
   $$x + (-x) = (a + b) + (-a, -b)$$
   $$= (0, 0) = 0.$$

   M1. (Closed under Multiplication) $xy = (ac - bd, ad + bc)$. Since ℝ is closed under addition and multiplication, $ac - bd \in \mathbb{R}$ and $ad + bc \in \mathbb{R}$, so indeed,
   $$(ac - bd, ad + bc) \in \mathbb{C}.$$

M2. (Commutative under Multiplication)
$$xy = (ac - bd, ad + bc)$$
$$= (ca - db, cb + da) = yx.$$

M3. (Associative under Multiplication)
$$(xy)z = (ac - bd, ad + bc)(e, f)$$
$$= (ace - bde - adf - bcf, acf - bdf + ade + bce)$$
$$= (a, b)(ce - df, cf + de) = x(yz).$$

M4. (Multiplicative Identity)
$$1x = (1, 0)(a, b)$$
$$= (a - 0, b + 0) = x.$$

M5. (Multiplicative Inverse)
$$(x)\frac{1}{x} = \left((a)\frac{a}{a^2 + b^2} - (b)\frac{-b}{a^2 + b^2}, (a)\frac{-b}{a^2 + b^2} + (b)\frac{a}{a^2 + b^2}\right)$$
$$= \left(\frac{a^2 + b^2}{a^2 + b^2}, \frac{-ab + ab}{a^2 + b^2}\right)$$
$$= (1, 0) = 1.$$

D. (Distributive Law)
$$x(y + z) = (a, b)(c + e, d + f)$$
$$= (ac + ae - bd - bf, ad + af + bc + be)$$
$$= (ac - bd, ad + bc) + (ae - bf, af + be) = xy + xz.$$
$\square$

For any $a, b \in \mathbb{R}$, we have $(a, 0) + (b, 0) = (a + b, 0)$ and $(a, 0)(b, 0) = (ab, 0)$. So a complex number of the form $(a, 0)$ behaves the same with respect to addition and multiplication as the corresponding real number a. If we associate $(a, 0) \in \mathbb{C}$ with $a \in \mathbb{R}$, we have the real field as a subfield of the complex field. (A *subfield* is a field that is also a subset of another field.)

The complex field is *not*, however, an ordered field. It fails axiom $O2$, since for $x = (0, 1)$, $x \neq 0$ but
$$x^2 = (0 - 1, 0 + 0)$$
$$= (-1, 0)$$
$$= (0 - 1, 0 - 0)$$
$$= 0 - 1$$
$$= -1 < 0.$$

Complex Numbers and Euclidean Spaces • 49

It turns out that our definition of complex numbers (a, b) is exactly the same as the more common $a + bi$, which uses the imaginary number $i = \sqrt{-1}$. Along with associating $a = (a, 0)$ for any real a, let us define $i = (0, 1)$. Then indeed,

$$i^2 = (0, 1)(0, 1)$$
$$= (-1, 0) = -1.$$

Furthermore, we have

$$a + bi = (a, 0) + (b, 0)(0, 1)$$
$$= (a, 0) + (0 - 0, b + 0) = (a, b).$$

Now we can understand the motivation behind the complicated (or should I say, *complex*?) multiplication rule: $(a, b)(c, d) = (ac - bd, ad + bc)$. If we multiply two complex numbers in the $a + bi$ form, we get

$$(a + bi)(c + di) = ac + adi + bci + bd(i^2)$$
$$= ac - bd + adi + bci = (ac - bd, ad + bc).$$

Definition 6.4. *(Complex Conjugate)*
For any $a, b \in \mathbb{R}$, let $z = a + bi$ (so $z \in \mathbb{C}$). We call a the **real part** of z, and we write $a = \text{Re}(z)$; we call b the **imaginary part** of z, and we write $b = \text{Im}(z)$.

Define $\bar{z} = a - bi$, so $\bar{z}$ is also a complex number, and we call it the **complex conjugate** (or just **conjugate**) of z.

Theorem 6.5. *(Properties of Conjugates)*
Let $z = a + bi$ and $w = c + di$ be complex numbers. The following properties hold:

Property 1. $\overline{z + w} = \bar{z} + \bar{w}$.
Property 2. $\overline{zw} = (\bar{z})(\bar{w})$.
Property 3. $z + \bar{z} = 2 \text{Re}(z)$, and $z - \bar{z} = 2i \text{Im}(z)$.
Property 4. $z\bar{z} \in \mathbb{R}$, and $z\bar{z} > 0$ (except when $z = 0$).

Proof. These should be a piece of cake. Notice that we only use the $a + bi$ form of the complex numbers, but the calculations are the same as they would be using the (a, b) form.

Property 1. Taking the conjugate distributes across addition, since

$$\overline{z + w} = \overline{(a + c) + (b + d)i}$$
$$= (a + c) - (b + d)i$$
$$= (a - bi) + (c - di) = \bar{z} + \bar{w}.$$

Property 2. Taking the conjugate distributes across multiplication, since

$$\overline{zw} = \overline{(ac - bd) + (ad + bc)i}$$
$$= (ac - bd) - (ad + bc)i$$
$$= (a - bi)(c - di) = (\bar{z})(\bar{w}).$$

Property 3. We have
$$z + \bar{z} = a + bi + (a - bi) = 2a = 2\,\text{Re}(z),$$
and similarly
$$z - \bar{z} = a + bi - (a - bi) = 2bi = 2i\,\text{Im}(z).$$

Property 4. We see that
$$z\bar{z} = (a + bi)(a - bi) = a^2 - (i^2)b^2 = a^2 + b^2,$$
which is a real number. Clearly it is positive as long as $z \neq 0$ (and it is 0 when $z = 0$). $\square$

Definition 6.6. *(Absolute Value)*
For any $z \in \mathbb{C}$, define the **absolute value** of z to be the positive square root of $z\bar{z}$.
  In symbols:
$$|z| = +(z\bar{z})^{\frac{1}{2}}.$$

The absolute value is, of course, always the *positive* square root of $z\bar{z}$. Since the previous theorem showed that $z\bar{z}$ is a real number ≥ 0, Theorem 5.8 guarantees the existence and uniqueness of any complex number's absolute value.

This definition of absolute value corresponds to the familiar one for real numbers. For any $x \in \mathbb{R}$, we have $x = \bar{x}$, so $|x| = +\sqrt{x^2}$. Thus

$$|x| = \begin{cases} x & \text{if } x \geq 0, \\ -x & \text{if } x < 0, \end{cases}$$

so that $|x| = \max\{x, -x\}$.

Theorem 6.7. *(Properties of Absolute Value)*
Let $z = a + bi$ and $w = c + di$ be complex numbers. Then the following properties hold:

  Property 1. $|z| > 0$ if $z \neq 0$, and $|z| = 0$ if $z = 0$.
  Property 2. $|\bar{z}| = |z|$.
  Property 3. $|zw| = |z|\,|w|$.
  Property 4. $|\text{Re}(z)| \leq |z|$, and $|\text{Im}(z)| \leq |z|$.
  Property 5. $|z + w| \leq |z| + |w|$.

Property 5 is called the *triangle inequality*. If you consider z, w, and $z + w$ as sides of a triangle, it asserts the main property of triangles—that any side is smaller than the sum of the other two sides.

Proof. Most of these proofs make use of the properties of conjugates we looked at in the last theorem.

  Property 1. $|z|$ is the positive square root of $z\bar{z}$, so it is positive as long as z is nonzero. If $z = 0$, then $|z| = (0\bar{0})^{\frac{1}{2}} = 0$.
  Property 2. In general,
$$\bar{\bar{z}} = \overline{a - bi}$$
$$= a + bi = z,$$

so we have
$$|\bar{z}| = (\bar{z}\bar{\bar{z}})^{\frac{1}{2}}$$
$$= (\bar{z}z)^{\frac{1}{2}} = |z|.$$

Property 3. By the field axiom $M2$ of $\mathbb{C}$,
$$|zw| = (zw\overline{zw})^{\frac{1}{2}} = (z\bar{z}w\bar{w})^{\frac{1}{2}}.$$

If $z = 0$ or $w = 0$, clearly $|zw| = 0 = |z||w|$. Otherwise, Property 4 of Theorem 6.5 tells us $z\bar{z}$ and $w\bar{w}$ are both real and positive, so by Corollary 5.9,
$$(z\bar{z}w\bar{w})^{\frac{1}{2}} = (z\bar{z})^{\frac{1}{2}}(w\bar{w})^{\frac{1}{2}},$$
and we have $|zw| = |z||w|$.

Property 4. We have
$$|\mathrm{Re}(z)| = |a|$$
$$= \sqrt{a^2}$$
$$\leq \sqrt{a^2 + b^2} \quad \text{(since } b^2 \geq 0\text{)}$$
$$= \sqrt{z\bar{z}} = |z|,$$
and similarly
$$|\mathrm{Im}(z)| = |b|$$
$$= \sqrt{b^2}$$
$$\leq \sqrt{a^2 + b^2} \quad \text{(since } a^2 \geq 0\text{)}$$
$$= \sqrt{z\bar{z}} = |z|.$$

Property 5. $\overline{\bar{z}w} = \bar{\bar{z}}\,\bar{w} = z\bar{w}$, so $\bar{z}w$ is the conjugate of $z\bar{w}$. By Property 3 of Theorem 6.5, this tells us $z\bar{w} + \bar{z}w = 2\mathrm{Re}(z\bar{w})$. We use that fact in the following inequality:
$$|z + w|^2 = (z+w)\overline{(z+w)}$$
$$= (z+w)(\bar{z} + \bar{w})$$
$$= z\bar{z} + z\bar{w} + \bar{z}w + w\bar{w}$$
$$= |z|^2 + 2\mathrm{Re}(z\bar{w}) + |w|^2$$
$$\leq |z|^2 + 2|z\bar{w}| + |w|^2 \quad \text{(by Property 4)}$$
$$= |z|^2 + 2|z||w| + |w|^2 \quad \text{(by Properties 2 and 3)}$$
$$= (|z| + |w|)^2.$$

Now since both sides of the inequality are real and ≥ 0, we can take square roots to obtain $|z + w| \leq |z| + |w|$. $\square$

Theorem 6.8. *(Cauchy-Schwarz Inequality)*
For any $a_1, a_2, \ldots, a_n \in \mathbb{C}$ and $b_1, b_2, \ldots, b_n \in \mathbb{C}$, the following inequality holds:

$$\left| \sum_{j=1}^{n} a_j \overline{b_j} \right|^2 \leq \left(\sum_{j=1}^{n} |a_j|^2 \right) \left(\sum_{j=1}^{n} |b_j|^2 \right).$$

What does this inequality even mean?! Sometimes summation signs can be confusing, and it's helpful to write them out as sums instead:

$$\left| a_1 \overline{b_1} + a_2 \overline{b_2} + \ldots + a_n \overline{b_n} \right|^2$$
$$\leq \left(|a_1|^2 + |a_2|^2 + \ldots + |a_n|^2 \right) \left(|b_1|^2 + |b_2|^2 + \ldots + |b_n|^2 \right).$$

Those conjugates in the left sum are a little confusing. We can simplify the inequality by substituting b_j for $\overline{b_j}$ (since the a_j and b_j's are arbitrary complex numbers), so we have

$$\left| a_1 \overline{\overline{b_1}} + a_2 \overline{\overline{b_2}} + \ldots + a_n \overline{\overline{b_n}} \right|^2$$
$$\leq (|a_1|^2 + |a_2|^2 + \ldots + |a_n|^2)(|\overline{b_1}|^2 + |\overline{b_2}|^2 + \ldots + |\overline{b_n}|^2),$$

which simplifies to

$$|a_1 b_1 + a_2 b_2 + \ldots + a_n b_n|^2$$
$$\leq \left(|a_1|^2 + |a_2|^2 + \ldots + |a_n|^2 \right) \left(|b_1|^2 + |b_2|^2 + \ldots + |b_n|^2 \right).$$

This looks kind of like a mutant triangle inequality, doesn't it? As we will soon see, one of its many applications is that it can be used to prove the triangle inequality for real vectors of any size.

Proof. Both sides of the Cauchy-Schwarz inequality are real numbers ≥ 0. If $(\sum_{j=1}^{n} |a_j|^2)(\sum_{j=1}^{n} |b_j|^2) = 0$, then it must be that $a_1 = a_2 = \ldots = a_n = 0$ and/or $b_1 = b_2 = \ldots = b_n = 0$, so clearly $|\sum_{j=1}^{n} a_j \overline{b_j}|^2$ also $= 0$ and we are done. Now we only need to prove the case in which both sides of the inequality are positive.

Since the sums range from 1 to some natural number n, we can use induction. First, we show the inequality holds for $\sum_{j=1}^{1}$; then we assume it holds for $\sum_{j=1}^{n-1}$ and prove it holds for $\sum_{j=1}^{n}$.

Base Case. For $n = 1$, we have

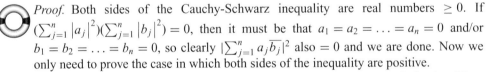

$$\left| \sum_{j=1}^{1} a_j \overline{b_j} \right|^2 = |a_1 \overline{b_1}|^2 = |a_1|^2 |b_1|^2 = \left(\sum_{j=1}^{1} |a_j|^2 \right) \left(\sum_{j=1}^{1} |b_j|^2 \right).$$

Inductive Step. The inductive hypothesis is

$$\left| \sum_{j=1}^{n-1} a_j \overline{b_j} \right|^2 \leq \sum_{j=1}^{n-1} |a_j|^2 \sum_{j=1}^{n-1} |b_j|^2.$$

Remember, we only need to worry about the case in which both sides are positive, so we can take the square root to obtain

$$\left| \sum_{j=1}^{n-1} a_j \overline{b_j} \right| \leq \sqrt{\sum_{j=1}^{n-1} |a_j|^2 \sum_{j=1}^{n-1} |b_j|^2}.$$

Thus

$$\left| \sum_{j=1}^{n} a_j \overline{b_j} \right| = \left| \left(\sum_{j=1}^{n-1} a_j \overline{b_j} \right) + a_n \overline{b_n} \right|$$

$$\leq \left| \sum_{j=1}^{n-1} a_j \overline{b_j} \right| + |a_n \overline{b_n}| \quad \text{(by the triangle inequality)}$$

$$\leq \sqrt{\sum_{j=1}^{n-1} |a_j|^2 \sum_{j=1}^{n-1} |b_j|^2} + |a_n \overline{b_n}| \quad \text{(by the inductive hypothesis)}$$

$$= \sqrt{\sum_{j=1}^{n-1} |a_j|^2} \sqrt{\sum_{j=1}^{n-1} |b_j|^2} + |a_n| \, |b_n|.$$

Here we're a little stuck. We want to be able to square $|a_n|$ and $|b_n|$ and bring them into their respective square-rooted sums. If we label

$$a = \sqrt{\sum_{j=1}^{n-1} |a_j|^2}, \, b = \sqrt{\sum_{j=1}^{n-1} |b_j|^2}, \, c = |a_n|, \text{ and } d = |b_n|,$$

we want to be able to say

$$ab + cd \leq \sqrt{a^2 + c^2} \sqrt{b^2 + d^2}.$$

In fact, we can say it! That inequality is always true for any $a, b, c, d \in \mathbb{R}$, because

$$0 \leq (ad - bc)^2 = a^2 d^2 - 2abcd + b^2 c^2$$
$$\implies 2abcd \leq a^2 d^2 + b^2 c^2$$
$$\implies a^2 b^2 + 2abcd + c^2 d^2 \leq a^2 b^2 + a^2 d^2 + b^2 c^2 + c^2 d^2$$
$$\implies (ab + cd)^2 \leq (a^2 + c^2)(b^2 + d^2),$$

and since both sides are positive reals, we can take the square root.

We use this inequality to obtain

$$\left|\sum_{j=1}^{n} a_j \overline{b_j}\right| \leq \sqrt{\sum_{j=1}^{n-1} |a_j|^2} \sqrt{\sum_{j=1}^{n-1} |b_j|^2} + |a_n||b_n|$$

$$\leq \sqrt{\sum_{j=1}^{n-1} |a_j|^2 + |a_n|^2} \sqrt{\sum_{j=1}^{n-1} |b_j|^2 + |b_n|^2}$$

$$= \sqrt{\left(\sum_{j=1}^{n} |a_j|^2\right)\left(\sum_{j=1}^{n} |b_j|^2\right)},$$

and just square both sides to complete the inductive step. $\square$

We can now generalize some ideas about complex numbers to real vectors of any size.

Definition 6.9. *(The Vector Space $\mathbb{R}^k$)*
*The set of all k-dimensional vectors of real numbers is denoted by $\mathbb{R}^k$. Each element—also called a **point**—of $\mathbb{R}^k$ is written as $\mathbf{x} = (x_1, x_2, \ldots, x_k)$, where $x_1, x_2, \ldots, x_k \in \mathbb{R}$ are called the **coordinates** of $\mathbf{x}$.*

*For $\mathbf{x}, \mathbf{y} \in \mathbb{R}^k$ and $\alpha \in \mathbb{R}$, define **vector addition** as*

$$\mathbf{x} + \mathbf{y} = (x_1 + y_1, x_2 + y_2, \ldots, x_k + y_k),$$

*and **scalar multiplication** as*

$$\alpha \mathbf{x} = (\alpha x_1, \alpha x_2, \ldots, \alpha x_k).$$

The vector $\mathbf{0} \in \mathbb{R}^k$ is defined as $\mathbf{0} = (0, 0, \ldots, 0)$.

*We call the set of real numbers $\mathbb{R}$ the **real line**, and the set of 2-dimensional real numbers $\mathbb{R}^2$ the **real plane**.*

Notice that we always have $\mathbf{x} + \mathbf{y} \in \mathbb{R}^k$ and $\alpha \mathbf{x} \in \mathbb{R}^k$, so $\mathbb{R}^k$ is closed under vector addition and scalar multiplication. The two operations are also associative, commutative, and distributive. Any set with operations that satisfies these conditions is called a *vector space*, another algebraic structure—*not* equivalent to a field—that is usually studied in linear algebra. Thus $\mathbb{R}^k$ is a vector space, over the real field. (But $\mathbb{R}^k$ is not a field, because there is no way to multiply two vectors.)

Definition 6.10. *(Euclidean Space)*
*For any $\mathbf{x}, \mathbf{y} \in \mathbb{R}^k$, define the **inner product** (or **scalar product**) of $\mathbf{x}$ and $\mathbf{y}$ as*

$$\mathbf{x} \cdot \mathbf{y} = \sum_{i=1}^{k} x_i y_i = x_1 y_1 + x_2 y_2 + \ldots + x_k y_k.$$

*Define the **norm** of $\mathbf{x}$ as*

$$|\mathbf{x}| = +\left(\sum_{i=1}^{k} x_i^2\right)^{\frac{1}{2}} = +\sqrt{x_1^2 + x_2^2 + \ldots + x_k^2}.$$

The vector space $\mathbb{R}^k$ together with the inner product and norm operations is a k-dimensional **Euclidean space**.

Distinction. Note that the scalar product is not the same as scalar multiplication. In both cases, *scalar* refers to a 1-dimensional vector (just a real number). The scalar product $\mathbf{x} \cdot \mathbf{y} \in \mathbb{R}$ is a way to multiply two vectors to yield a scalar; scalar multiplication $\alpha \mathbf{x} \in \mathbb{R}^k$ is a way to multiply a scalar by a vector to yield a vector.

Also, for real 2-vectors, the scalar product is not the same as the complex product. For complex numbers $z = (a, b)$ and $w = (c, d)$, $z \cdot w = ac + bd$ is the scalar product, whereas $zw = (ac - bd, ad + bc)$ is the complex product. Note that we have not defined a cognate of the complex product for real vectors of dimension higher than 2, so $\mathbb{R}^k$ is *not* a field (we would need a way to multiply two vectors and get a *vector*).

The norm is, of course, always the *positive* square root of $\mathbf{x} \cdot \mathbf{x}$. Since the scalar product of $\mathbf{x}$ with itself is always a real number, Theorem 5.8 guarantees the existence and uniqueness of any real vector's norm.

Remember that the absolute value of $(a, b) \in \mathbb{C}$ is defined as

$$|(a, b)| = \sqrt{(a, b)(a, -b)} = \sqrt{a^2 + b^2},$$

So the norm of a complex number or a real number is just its absolute value.

Theorem 6.11. *(Properties of the Norm)*
Let $\mathbf{x}, \mathbf{y}, \mathbf{z} \in \mathbb{R}^k$, and let $\alpha \in \mathbb{R}$. Then the following properties hold:

> *Property 1.* $|\mathbf{x}| > 0$ if $\mathbf{x} \neq \mathbf{0}$, and $|\mathbf{x}| = 0$ if $\mathbf{x} = \mathbf{0}$.
> *Property 2.* $|\alpha \mathbf{x}| = |\alpha|\,|\mathbf{x}|$.
> *Property 3.* $|\mathbf{x} \cdot \mathbf{y}| \leq |\mathbf{x}|\,|\mathbf{y}|$.
> *Property 4.* $|\mathbf{x} + \mathbf{y}| \leq |\mathbf{x}| + |\mathbf{y}|$.
> *Property 5.* $|\mathbf{x} - \mathbf{z}| \leq |\mathbf{x} - \mathbf{y}| + |\mathbf{y} - \mathbf{z}|$.
> *Property 6.* $|\mathbf{x} - \mathbf{y}| \geq |\mathbf{x}| - |\mathbf{y}|$.

Although this theorem tells us that $|\mathbf{x}| \geq 0$, note that it doesn't make sense to write something like "$\mathbf{x} \geq \mathbf{0}$," since no order is defined for the elements in $\mathbb{R}^k$.

Property 4 asserts the triangle inequality for Euclidean spaces. (Actually, Property 5 is more deserving of the name "triangle inequality," because if we treat $\mathbf{x}, \mathbf{y}$, and $\mathbf{z}$ as the three points of a triangle, $|\mathbf{x} - \mathbf{z}|$, $|\mathbf{x} - \mathbf{y}|$, and $|\mathbf{y} - \mathbf{z}|$ represent the lengths of its three sides.)

Proof. Most of these properties are similar, but not identical, to their cognates for absolute value in Theorem 6.7.

> *Property 1.* $|\mathbf{x}|$ is the positive square root of $\mathbf{x} \cdot \mathbf{x}$, so it is positive as long as $\mathbf{x}$ is nonzero. If $\mathbf{x} = \mathbf{0}$, then
>
> $$|\mathbf{0}| = (\mathbf{0} \cdot \mathbf{0})^{\frac{1}{2}} = \sqrt{0 + 0 + \ldots + 0} = 0.$$

Property 2. We have
$$|\alpha\mathbf{x}| = \sqrt{\alpha\mathbf{x} \cdot \alpha\mathbf{x}}$$
$$= \sqrt{\alpha^2 x_1^2 + \alpha^2 x_2^2 + \ldots + \alpha^2 x_k^2}$$
$$= \sqrt{\alpha^2}\sqrt{\mathbf{x} \cdot \mathbf{x}} = |\alpha|\,|\mathbf{x}|.$$

Property 3. Let
$$a_1 = x_1, a_2 = x_2, \ldots, a_k = x_k, \text{ and } \overline{b_1} = y_1, \overline{b_2} = y_2, \ldots, \overline{b_k} = y_k.$$

Since the coordinates of $\mathbf{x}$ and $\mathbf{y}$ are real numbers—which are just complex numbers of the form $(a, 0)$—we know that every a_i and b_i is a complex number.

Then by the Cauchy-Schwarz inequality,
$$|\mathbf{x} \cdot \mathbf{y}|^2 = \left|\sum_{i=1}^{k} x_i y_i\right|^2$$
$$= \left|\sum_{i=1}^{k} a_i \overline{b_i}\right|^2$$
$$\leq \left(\sum_{i=1}^{k} |a_i|^2\right)\left(\sum_{i=1}^{k} |b_i|^2\right)$$
$$= \left(\sum_{i=1}^{k} |x_i|^2\right)\left(\sum_{i=1}^{k} |y_i|^2\right)$$
$$= |\mathbf{x}|^2 |\mathbf{y}|^2.$$

Since both sides are ≥ 0, we can take the square root.

Property 4. Using the previous property, we have
$$|\mathbf{x} + \mathbf{y}|^2 = (\mathbf{x} \cdot \mathbf{y}) \cdot (\mathbf{x} \cdot \mathbf{y})$$
$$= \sum_{i=1}^{k}(x_i + y_i)(x_i + y_i)$$
$$= \sum_{i=1}^{k}(x_i^2 + 2x_i y_i + y_i^2)$$
$$= \mathbf{x} \cdot \mathbf{x} + 2\mathbf{x} \cdot \mathbf{y} + \mathbf{y} \cdot \mathbf{y}$$
$$\leq |\mathbf{x} \cdot \mathbf{x}| + 2|\mathbf{x} \cdot \mathbf{y}| + |\mathbf{y} \cdot \mathbf{y}|$$
$$\leq |\mathbf{x}||\mathbf{x}| + 2|\mathbf{x}||\mathbf{y}| + |\mathbf{y}| + |\mathbf{y}|$$
$$= (|\mathbf{x}| + |\mathbf{y}|)^2.$$

Because both sides are ≥ 0, we can take the square root.

Property 5. This is actually the same as the previous property; just substitute $\mathbf{x} - \mathbf{y}$ for $\mathbf{x}$, and $\mathbf{y} - \mathbf{z}$ for $\mathbf{y}$.

Property 6. In the previous property, set $\mathbf{z} = \mathbf{0}$ and subtract $|\mathbf{y}|$ from both sides.

□

Why bother with imaginary numbers? As you learned, the complex numbers are just $\mathbb{R}^2$ with special addition and multiplication tacked on. Learning to manipulate new sets, fields, and vector spaces is an important part of real analysis.

Soon we'll begin our study of topology by introducing another important concept in set theory: functions and an extra-special type of function called a *bijection*.

TOPOLOGY

CHAPTER 7

Bijections

At a basic level, the field of *topology* is concerned with the properties of sets and their elements. We'll learn a host of topological definitions in Chapter 9, but first we want to understand the notion of *countability*. To do so, we need to spend this chapter going over some important properties that functions can have, with the goal of building toward the definition of a *bijection*.

I'm sure you just let out a groan when you read the word *functions*. You've probably seen them over and over, defined and redefined throughout years of math classes. But we still need to define functions formally and make them consistent with our notation for sets. You can't have calculus—and thus real analysis—without functions.

Definition 7.1. *(Function)*
*If, for each element x in the set A, there is a corresponding unique element $f(x)$ in the set B, then the relation f is a **function** (also called a **mapping**). We say f **maps** A to B, and we write*

$$f: A \to B, \quad f: x \mapsto f(x).$$

*The set A is called the **domain** of f, and the set B is called the **codomain** of f.*

Example 7.2. *(Functions)*
Figure 7.1 demonstrates a relation that is not a function.
 The following is a function with domain $A = \mathbb{R}$ and codomain $B = \mathbb{R}$:

$$f: \mathbb{R} \to \mathbb{R}, \quad f: x \mapsto x^2.$$

For short, we can write this function as $f(x) = x^2$. This does leave room for some ambiguity, though, since we have not specified which sets are being mapped—we might have $f: \mathbb{Q} \to \mathbb{Q}$, or even $f: \{1, 2, 3\} \to \{1, 4, 9\}$. Whenever it is necessary to know which sets are being mapped, we will write it out explicitly.
 Another example is

$$f: x \mapsto 1, \quad \forall x \in \mathbb{N},$$

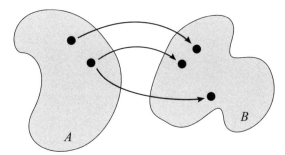

Figure 7.1. The pictured relation is *not* a function $A \to B$, since it maps the same element of A to two different elements of B.

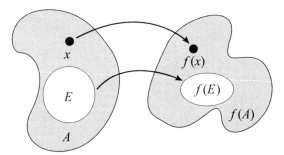

Figure 7.2. The function f acting on the set A. Its range is $f(A)$. It maps the element x to the element $f(x)$ and the set E to the set $f(E)$—both of which are, of course, contained in the range $f(A)$.

which is a function with domain $A = \mathbb{N}$. Here, the codomain is not specified.

On the other hand,

$$f : x \mapsto \sqrt{x}, \quad \forall x \in \mathbb{R}$$

is not a function, since the same x is mapped to multiple values—namely, to $+\sqrt{x}$ and $-\sqrt{x}$. Note that both $f(x) = +\sqrt{x}$ and $g(x) = -\sqrt{x}$ *are* functions.

Definition 7.3. *(Image and Range)*
For any function $f : A \to B$ and any $E \subset A$, $f(E)$ is the set of all elements in E that can be mapped to by f from some element of A. The set $f(E)$ is called the **image** of E under f.

In symbols, the image of E under f is the set:

$$f(E) = \{f(x) \mid x \in E\}.$$

The set $f(A)$ is called the **range** of f.

Example 7.4. (Images and Ranges)
Figure 7.2 demonstrates the image of a set and the range of a function.

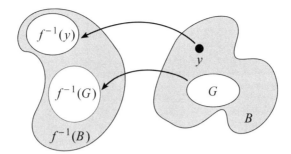

Figure 7.3. The inverse image of G, B, and $\{y\}$ under the function f.

If $E = (-3, 3)$ and

$$f: \mathbb{R} \to \mathbb{R}, \quad f: x \mapsto x^2,$$

then the image of E under f is the half-open interval $f(E) = [0, 9)$. The range of f is $[0, \infty)$, because every positive real number is the square of some real number.

If $E = \{1, 2, 3\}$ and

$$f: x \mapsto 1, \quad \forall x \in \mathbb{N},$$

then the image of E under f is the set $f(E) = \{1\}$. In fact, the range of f is also the set $\{1\}$.

Definition 7.5. *(Inverse Image)*
For any function $f: A \to B$ and any $G \subset B$, $f^{-1}(G)$ is the set of all elements of A whose image under f is contained in G. The set $f^{-1}(G)$ is called the **inverse image** of G under f.

In symbols, the inverse image of G under f is the set:

$$f^{-1}(G) = \{x \in A \mid f(x) \in G\}.$$

For any $y \in B$, $f^{-1}(y)$ is the set of all elements of A whose image is y.
In symbols, for any $y \in B$:

$$f^{-1}(y) = \{x \in A \mid f(x) = y\}.$$

Example 7.6. *(Inverse Images)*
Figure 7.3 demonstrates a function f and the inverse image of a few sets.

The function $f(x) = x^2$, $\forall x \in \mathbb{Z}$ maps $\mathbb{Z} \to \mathbb{N} \cup \{0\}$. The inverse image under f of the set $\mathbb{N}$ is the set of all integers whose square is a natural number, so $f^{-1}(\mathbb{N}) = \mathbb{Z} \setminus \{0\}$. The inverse image under f of the set containing the single number 16 is $f^{-1}(\{16\}) = \{4, -4\}$. Note that this implies f^{-1} is not a function (since it maps a single element to two different elements).

Definition 7.7. *(Inverse)*
For any function $f: A \to B$, if the relation $f^{-1}: B \to A$ is a function, then we call f^{-1} the **inverse** of f. If such an inverse exists, then f is said to be **invertible**.

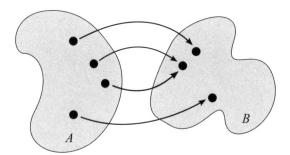

Figure 7.4. If B contains only the three pictured points, then this function is onto/surjective (since every element of B is mapped to by at least one element of A).

Example 7.8. (Inverses)
The function

$$f: \mathbb{R} \to \mathbb{R}, \quad f: x \mapsto x^2$$

is not invertible, since its inverse $f^{-1}(y) = \sqrt{x}$ is not a function (because it maps to both $+\sqrt{x}$ and $-\sqrt{x}$).

On the other hand, the function

$$f: x \mapsto 2x, \quad \forall x \in \mathbb{R}$$

is invertible, since its inverse is $f^{-1}(y) = \frac{y}{2}$.

 Note that we chose to write $f^{-1}(y) = \frac{y}{2}$ instead of $f^{-1}(x) = \frac{x}{2}$. Although they mean the same thing, we want to be clear that for $f: A \to B$, f^{-1} acts on elements y of B, not on elements x of A.

In the previous example, if we restricted the domain of f to be the set $\mathbb{N}$, then f would not be invertible, since $f^{-1}(3) = \frac{3}{2}$ is not a natural number.

To understand which functions are invertible, we need to learn about two special types of functions: *onto* and *one-to-one*.

Definition 7.9. *(Onto Function)*
*For any function $f: A \to B$, if $f^{-1}(y)$ contains at least one element of A for every $y \in B$, then f is an **onto** mapping of A into B. An onto function is also called a **surjective** function or a **surjection**.*

A function is surjective if every element of the codomain B can be mapped to by some element of the domain A. In other words, the range of f must be all of B, meaning $f(A) = B$.

Example 7.10. (Onto Functions)
Figure 7.4 demonstrates an onto function.
If $B = \{1\}$, then the function

$$f: \mathbb{N} \to B, \quad f: x \mapsto 1$$

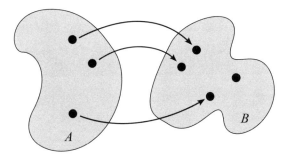

Figure 7.5. If A contains only the three pictured points, then this function is one-to-one/injective (since every element of B is mapped to by at most one element of A).

is onto. But if $B \neq \{1\}$, then f is not onto, since then there are elements of B that have no inverse image under f.

Any function that is not onto cannot be invertible, since f^{-1} is not defined for all elements of B.

The converse, however, is not necessarily true: there do exist functions that are not invertible but *are* onto. For example, the function

$$f: \{1, 2\} \to \{1\}, \quad f: x \mapsto 1,$$

is onto, since every element of B (i.e., its only element) is mapped to by f. But f is not invertible, since $f^{-1}(1)$ maps to two different elements of A.

Definition 7.11. *(One-to-One Function)*
For any function $f: A \to B$, if $f^{-1}(y)$ contains at most *one element of A for every $y \in B$, then f is a **one-to-one** mapping of A into B. A one-to-one function is also called an **injective** function or an **injection**.*

A function is injective if every element of the codomain B is mapped to by no more than one element of the domain A. In other words, for every pair of distinct elements $x_1, x_2 \in A$ (*distinct* meaning $x_1 \neq x_2$), we must have $f(x_1) \neq f(x_2)$.

Example 7.12. (One-to-One Functions)
Figure 7.5 demonstrates a one-to-one function.
 The function

$$f: x \mapsto 2x, \quad \forall x \in \mathbb{R}$$

is one-to-one, since $x_1 \neq x_2 \implies 2x_1 \neq 2x_2$.
 On the other hand, the function

$$f: x \mapsto x^2, \quad \forall x \in \mathbb{R}$$

is not one-to-one, because, for example, the number 4 has two inverse images: -2 and 2.

As with a function that is not a surjection, any function that is not an injection is also not invertible: if some element of B is mapped to by at least two elements of A, then f^{-1} is not a function.

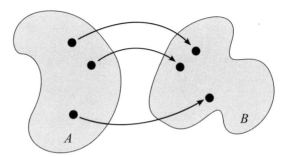

Figure 7.6. If A and B each contain only the three pictured points, then this function is a bijection (since every element of B is mapped to by exactly one element of A).

The converse, again, is not necessarily true: there do exist functions that are not invertible but *are* injective. For example, the function

$$f: \{1\} \to \{1, 2\}, \quad f: x \mapsto 1,$$

is injective, since no element of B is mapped to by more than one element of A. But f is not invertible, since $f^{-1}(2)$ is not defined.

Definition 7.13. *(Bijection)*
A **bijection** is a function that is both onto and one-to-one.

Example 7.14. (Bijections)
Figure 7.6 demonstrates a bijection. Contrast it to Figure 7.4 and Figure 7.5, neither of which exhibit a bijection.

Out of all the functions we have seen in examples so far, only

$$f: x \mapsto 2x, \quad \forall x \in \mathbb{R}$$

is a bijection.

Theorem 7.15. *(Bijection $\iff$ Invertible)*
A function $f: A \to B$ is a bijection if and only if it is invertible.

Proof. Let's start with showing that a bijection is invertible. For f^{-1} to exist, it must be defined for every element of B, and each element of B can only be mapped to one element of A (if it were mapped to two different elements of A, f^{-1} wouldn't be a function). This proof is just hiding in the definitions! Since f is surjective, every element of B is mapped to from some element of A, so f^{-1} is defined for every element of B. Since f is injective, no two elements of B are mapped to from the same element of A, so each value of $f(y)$ is unique $\forall y \in B$.

The other direction is very similar. If f is invertible, then f^{-1} is defined for every element of B, so f must map to every element of B from some elements of A, so f is surjective. Also, f^{-1} is a function, so each element $x \in A$ has no more than one $f(x) \in B$, so f is injective.

□

Actually, not only does the inverse of a bijection *exist*, it's also always a bijection itself!

Theorem 7.16. *(Inverse of a Bijection)*
If $f: A \to B$ is a bijection, then $f^{-1}: B \to A$ is also a bijection.

Proof. You can do this one! Fill in the blanks in Box 7.1.

BOX 7.1

> PROVING THEOREM 7.16
>
> Since f is injective, each element of B is mapped to by f from at most one element of A, so f^{-1} is a _____. Since f must be defined for every element of ____, the set A is the range of ____.
>
> Since f is surjective, every element of B is mapped to by f from at _____ one element of A, so ____ is the domain of f^{-1}.
>
> Since f is a function, each element of A can only map to one element of B, so f^{-1} is _____.
>
> Thus f^{-1} is an injective, surjective function, so it is a _____.

□

Definition 7.17. *(Function Composition)*
The **composition** of two functions $f: A \to B$ and $g: B \to C$, is the function
$$g \circ f: A \to C, \; g \circ f: x \mapsto g(f(x)).$$

Example 7.18. (Function Compositions)
Compositions are pretty straightforward; you just apply one function and then the other. If we have $f(x) = 2x$ and $g(x) = x^2$, then
$$(g \circ f)(x) = g(f(x)) = (2x)^2 = 4x^2.$$

Always remember that to compose two functions $f: A \to B$ and $g: B \to C$, the codomain B of f must be the same as the domain B of g. If we want to reverse the order of composition to calculate $f \circ g$, we must have $C = A$.

Note that the composition of a function and its inverse function (in either order) is just the identity mapping $x \mapsto x$, since
$$(f^{-1} \circ f)(x) = f^{-1}(f(x)) = x = f(f^{-1}(x)) = (f \circ f^{-1})(x).$$

In the next chapter, bijections will be used in trying to understand the meaning of an infinite set better. There is more than one type of infinity, as you may have heard before, and that makes our lives more complicated (but more *fun* too, amiright?!).

CHAPTER 8

Countability

As promised, we'll take a look at two different types of infinite sets: *countable* and *uncountable*. We'll explore examples of each, and we'll learn that $\mathbb{Q}$ is countable while $\mathbb{R}$ is not. These definitions are rooted in the measure of *cardinality*, which is a type of *equivalence relation*.

Definition 8.1. *(Equivalence Relation)*
A relation $\equiv$ between two objects a and b is called an **equivalence relation** if it satisfies the following three properties:

  Property 1. *(Reflexive)* $a \equiv a$.
  Property 2. *(Symmetric)* If $a \equiv b$, then $b \equiv a$.
  Property 3. *(Transitive)* If $a \equiv b$ and $b \equiv c$, then $a \equiv c$.

In this definition, $\equiv$ is used as a stand-in for the symbol of any arbitrary relation.

Example 8.2. *(Equivalence Relations)*
Clearly, the relation of equality, denoted $=$, is an equivalence relation for numbers and elements of sets. It is also an equivalence relation for sets themselves, since for any set A, we have

$$A = A; \quad A = B \implies B = A; \quad \text{and} \quad A = B, B = C \implies A = C.$$

Definition 8.3. *(Cardinality)*
Let us define the relation $\sim$ as follows: For any two sets A and B, $A \sim B$ if and only if there exists a bijection $f : A \to B$.
  If $A \sim B$, then we say that A and B can be put into **one-to-one correspondence**, and that A and B have the same **cardinal number** (or **cardinality** for short).

Distinction. The phrase "one-to-one correspondence" is not the same as a one-to-one function. One-to-one correspondence is used in this case for two sets that are related by a function that is both one-to-one *and* onto.

Theorem 8.4. *(Cardinality Is an Equivalence Relation)*
The relation $\sim$ is an equivalence relation.

Proof. For each of the three properties of equivalence relations, we need to show that there exists a bijection f that satisfies that property.

> *Property 1.* $A \sim A$.
> We need a function f that maps any set A to itself bijectively (is that a word? I don't think so). What do you guess the $A \to A$ bijection is?
> I guess it's $f: x \mapsto x$. (And I'm probably right, since I wrote this book.) Let's check: f maps every element in A to the same (single) element in A, so every element of the codomain is mapped by f from a single element of the domain, so f is a bijection that maps $A \to A$.
>
> *Property 2.* $A \sim B \implies B \sim A$.
> Assume f is a bijection $A \to B$, so we need a bijection $B \to A$.
> Remember Theorem 7.16? It tells us that f^{-1} is exactly what we're looking for.
>
> *Property 3.* $A \sim B, B \sim C \implies A \sim C$.
> Let f be the $A \to B$ bijection, and let g be the $B \to C$ bijection.
> In general, the composition of two bijections is also a bijection, since f and g are both injective, so each element in C comes from at *most* one element in B (via g), which comes from at most one element in A (via f), so $g \circ f$ is also injective. Also, f and g are both surjective, so each element in C comes from at *least* one element in B (via g), which comes from at least one element in A (via g), so $g \circ f$ is also surjective.
> Thus $g \circ f$ is the bijection we need. □

Wait. I thought the statement "A equals B" means that A and B have the exact same elements, not that there exists a bijection between them! Remember the difference between equivalence and equality: equality is one *type* of equivalence relation between sets, and $\sim$ is another type of equivalence relation.

It turns out that for finite sets, two sets with the same cardinality (i.e., being $\sim$ equivalent) means they have the same number of elements.

Theorem 8.5. *(Cardinality for Finite Sets)*
Let A and B be finite sets. Then $A \sim B$ if and only if A and B have the same number of elements.

Proof. For finite sets A and B, we will prove both directions of the "if and only if."

If $A \sim B$, then there exists a function that maps *every* element of A (since f is a well-defined function) to at most one element of B (since f is injective), and every element of B is mapped to by some element of A (since f is surjective). So for every element of A there is a corresponding element of B, and all elements of B are covered by this correspondence, so A and B must have the same number of elements.

If A and B have the same number n of elements, then we can write the sets as $A = \{a_1, a_2, \ldots, a_n\}$ and $B = \{b_1, b_2, \ldots, b_n\}$. Let $f: A \to B$ be defined as $f: a_i \mapsto b_i$, for every $1 \leq i \leq n$. Then f is a one-to-one, onto function, so f is a bijection. We have found a bijection $A \to B$, so $A \sim B$. □

As we will soon see, this interpretation of one-to-one correspondence also (sort of) translates to infinite sets. An infinite set cannot have the same "number" of elements as another infinite set (since the number of elements they both have is infinite), but they can have the same "type of infinity" number of elements.

Definition 8.6. *(Countability)*
*For any set A, we say A is **countable** if $A \sim \mathbb{N}$. A is **at most countable** if A is either finite or countable. A is **uncountable** if A is both infinite and not countable.*

To help us understand this definition, let's reconsider what it means for a set to be finite. We know A has a finite number of elements if and only if we can write them as $A = \{a_1, a_2, \ldots, a_n\}$ for some finite $n \in \mathbb{N}$. Really, this means that there is a bijection between the set of numbers $\{1, 2, \ldots, n\}$ and the set A. This bijection looks like

$$f: 1 \mapsto a_1, f: 2 \mapsto a_2, \ldots, f: n \mapsto a_n.$$

So

$$A \text{ is finite} \iff A \sim \mathbb{N}_n.$$

(Here $\mathbb{N}_n$ means the subset of the natural numbers constituting the first n numbers.)

Simply put, A being infinite but countable means you can "count" its elements with the natural numbers. The elements of A can be laid out in some pattern, and this pattern can be mapped via a bijection to the natural numbers.

Some infinite sets are countable, and others are not—just like some finite sets have four elements, and others have three. It turns out that two sets with the same cardinality means they both have the same "type" of number of elements (finite, countably infinite, or uncountably infinite).

Theorem 8.7. *(Cardinality and Countability)*
For any two sets A and B, if $A \sim B$, then either A and B are finite with the same number of elements, or A and B are both countable, or A and B are both uncountable.

Proof. We already proved the finite case (if A, B finite, then $A \sim B \iff A$ and B have the same number of elements) in Theorem 8.5, so we only need to consider the case in which A and B are infinite.

Assume $A \sim B$. Then if A is countable, $A \sim \mathbb{N}$, so by Property 2 of Theorem 8.4, $\mathbb{N} \sim A$, so by Property 3 of Theorem 8.4, $\mathbb{N} \sim B$, so B is also countable.

If A is uncountable, then B must be as well. Why? Assume B is countable, so $B \sim \mathbb{N}$. Then $\mathbb{N} \sim B$, so $\mathbb{N} \sim A$, which is a contradiction (we said A is uncountable). □

Of course, the converse of this theorem is only partially true. If A and B are both countable, then $A \sim \mathbb{N}$ and $B \sim \mathbb{N}$, so $A \sim B$. But if A and B are both uncountable, all we have is $A \sim ?$, where ? is *not* $\mathbb{N}$, but this tells us nothing about whether A and B can be put in one-to-one correspondence. B could have an infinitely larger number of infinite elements than A.

What you should take away from this theorem is an understanding of what cardinality really means: it is a measure of whether a set is finite, countable, or uncountable.

Example 8.8. ($\mathbb{Z}$ Is Countable)
If we can count forward, then we can also count backward—I claim that the set of all integers, $\mathbb{Z}$, is countable. We want to show that there exists a bijection $f: \mathbb{N} \to \mathbb{Z}$. (By Theorem 7.16, we could instead find one that maps the other way, but this direction will be easier.)

However, we cannot just say, "map $\mathbb{N}$ to count forward to infinity, and map it again to count backward to negative infinity, to cover all integers"—that would be a one-to-two mapping (since each natural number is mapped to itself and its negative), which is not a function.

Instead, we may just alternate between each integer as follows:

$$f: 2 \mapsto 1, f: 3 \mapsto -1, f: 4 \mapsto 2, f: 5 \mapsto -2, \ldots$$

If we are able to formalize the definition of this function, then we can easily prove it is a bijection.

$$\begin{array}{cccccccc} 1 & 2 & 3 & 4 & 5 & 6 & 7 & \ldots \\ \downarrow & \downarrow & \downarrow & \downarrow & \downarrow & \downarrow & \downarrow & \downarrow \\ 0 & 1 & -1 & 2 & -2 & 3 & -3 & \ldots \end{array}$$

For every $n \in \mathbb{N}$, let

$$f(n) = \begin{cases} \frac{n}{2} & \text{if } n \text{ is even,} \\ -\frac{n-1}{2} & \text{if } n \text{ is odd.} \end{cases}$$

The function f is defined for every $n \in \mathbb{N}$, and every natural number is mapped to exactly one integer, so $f: \mathbb{N} \to \mathbb{Z}$ is a well-defined function. No two natural numbers map to the same integer, so f is injective. Every integer can be written either as $\frac{n}{2}$ for some even n, or as $-\frac{n-1}{2}$ for some odd n, so f is surjective. Thus f is indeed a bijection.

Notice that in this example, we showed $\mathbb{N} \sim \mathbb{Z}$, even though $\mathbb{N}$ is a *proper* subset of $\mathbb{Z}$. This proper-but-equivalent relationship is possible because both sets are infinite. If either were finite, then by Theorem 8.5, they would both need to have the same number of elements to have the same cardinality, so one could not be a proper subset of the other.

In fact, the formal definition of an infinite set is as follows.

Definition 8.9. *(Infinite Set)*
An **infinite set** *is one that has the same cardinality as at least one of its proper subsets.*

Of course, under the equivalence relation $=$, no set can be equivalent to one of its proper subsets (even if both are infinite), since Definition 3.5 defines a proper subset as a nonequal subset.

Example 8.10. (An Uncountable Set)
Here is what you've been waiting for: an example of an uncountable set. We're going to look at the set S of all real numbers between 0 and 1, which by Definition 3.8 we can write as $(0, 1)$.

How are we going to prove S is uncountable? Well if S is uncountable, then if we take a countable subset E of S, there should be some elements that are in S but not in E, right?

So, if we can show that *every* countable subset of S is a *proper* subset, then S is uncountable. Because if every countable subset of S is a proper subset but S were countable, then S itself would be a proper subset of itself, which is impossible.

Basically what all this means is that no matter how we try to count the elements of S, some will always be left out of the counting.

We will take for granted the theorem from number theory which says that every real number has an infinite decimal expansion—so we can write every number in S as $0.d^1 d^2 d^3 d^4 \ldots$ where each decimal d^i is an integer between 0 and 9, inclusive.

In this case, d^i does not mean d raised to the ith power; it means the ith element of the set $\{d^1, d^2, d^3, \ldots\}$. This ambiguous notation is definitely confusing, but unfortunately it's used all the time, so you better get used it. (You can usually tell from the context when a superscript represents a power and when it represents an index.)

Note that some rational numbers have finite decimal expansions, but we can tack on an infinite string of zeros at the end.

Now, take a countable subset $E \subset S$. The elements of E can be numbered $e_1, e_2, e_3, e_4, \ldots$, so we can arrange them in order.

$$e_1 = 0.\ d_1^1\ d_1^2\ d_1^3\ d_1^4\ \ldots$$
$$e_2 = 0.\ d_2^1\ d_2^2\ d_2^3\ d_2^4\ \ldots$$
$$e_3 = 0.\ d_3^1\ d_3^2\ d_3^3\ d_3^4\ \ldots$$
$$e_4 = 0.\ d_4^1\ d_4^2\ d_4^3\ d_4^4\ \ldots$$
$$\vdots$$

Let's form the element $s = 0.s^1 s^2 s^3 s^4 \ldots$ as follows. If d_1^1 (the first decimal of e_1) is 0, let s^1 be 1. Otherwise, let $s^1 = 0$. Then $s^1 \neq d_1^1$, so s cannot equal e_1. Repeat this pattern for every decimal of s. To make this more precise, we define the following rule for every $i \in \mathbb{N}$:

$$s^i = \begin{cases} 1 & \text{if } d_i^i = 0, \\ 0 & \text{if } d_i^i \neq 0. \end{cases}$$

For example, if we have

$$e_1 = 0.\ 3\ 2\ 0\ 8\ \ldots$$
$$e_2 = 0.\ 9\ 0\ 6\ 6\ \ldots$$
$$e_3 = 0.\ 1\ 5\ 0\ 7\ \ldots$$
$$e_4 = 0.\ 2\ 4\ 2\ 7\ \ldots$$
$$\vdots$$

then $s = 0.0110\ldots$.

Thus, s differs from each e_i by at least one decimal—specifically, the ith decimal—so clearly, s cannot equal e_i for any $i \in \mathbb{N}$. Remember that since E is countable, every element of E can be written as e_i for some $i \in \mathbb{N}$, so we have just shown that $s \notin E$. But $s \in S$, so E must be a proper subset of S.

E was arbitrary, so we just proved that every countable subset of S is a proper subset. Thus S is uncountable.

This argument is called the *Cantor diagonal process*, since Georg Cantor was the first person to do this. Why is it called a "diagonal process"? Remember how we constructed that extra element s by making it differ from one decimal of each e_i? Well, s differs from each e_i on the ith decimal, which, if you notice, makes s differ from the set E by each decimal on the diagonal of the chart we used:

$$
\begin{aligned}
e_1 &= 0 \,.\, d_1^1 \; d_1^2 \; d_1^3 \; d_1^4 \; \ldots \\
e_2 &= 0 \,.\, d_2^1 \; d_2^2 \; d_2^3 \; d_2^4 \; \ldots \\
e_3 &= 0 \,.\, d_3^1 \; d_3^2 \; d_3^3 \; d_3^4 \; \ldots \\
e_4 &= 0 \,.\, d_4^1 \; d_4^2 \; d_4^3 \; d_4^4 \; \ldots \\
&\vdots
\end{aligned}
$$

The previous example serves us for two reasons. First, when combined with the upcoming theorem, it will show that $\mathbb{R}$ is uncountable.

Second, it gives us a better sense of what it means to be uncountable. The set S failed because we could *not* arrange its elements in an order. For any pattern we try, there will always be some elements left out of the pattern. (Because if we *could* find a pattern that included all the elements, we could find a bijection mapping the natural numbers to that pattern, and so the set would be countable.)

Theorem 8.11. *(Cardinality of Subsets and Supersets)*
Let E be a subset of A. If A is at most countable, then E is also at most countable. If E is uncountable, then A is also uncountable.

Proof. The first few cases are simple. Let A be at most countable, so A is either finite or countably infinite. If A is finite, then clearly E is finite as well, so E is at most countable. If A is countably infinite, then either E is finite or infinite. If E is finite, then again E is at most countable.

The crux of the proof thus comes down to the case in which A is countably infinite and E is infinite. Since $A \sim \mathbb{N}$, we can arrange its elements in some order $\{x_1, x_2, x_3, \ldots\}$. Clearly, some of the elements in this sequence are in E and others are not. Then our goal is to use this ordering we have for A's elements and apply it to E's elements. We cannot just map the first element of A to the first element of E and so on, since each element in A might not also be in E. Instead, to come up with an ordering for, say, 10 elements of E, we could just say, "look at the first 10 elements of A that are in E, and make those the first 10 elements of E, in the same order."

Let's formalize this. Let n_1 be the smallest natural number such that x_{n_1} is in E. Let n_2 be the smallest natural number greater than n_1, such that x_{n_2} is in E. Having chosen $k - 1$ elements in this sequence of indexes, let the next element n_k be the smallest natural number greater than n_{k-1}, such that x_{n_k} is in E. (For any $k \in \mathbb{N}$, n_k will always exist, since the sequence $\{x_i\}$ of elements of A is infinite, and an infinite number of these elements are in the subset E.) Then we have a nice bijection $f : \mathbb{N} \to \mathbb{N}$, $f : k \mapsto n_k$ for any $k \in \mathbb{N}$.

Now we can write the elements of E as those elements of A which are also in E, so $E = \{x_{n_1}, x_{n_2}, x_{n_3}, \ldots\}$.

This notation is used to describe a *subsequence*. We'll get a lot of practice with subsequences in Chapter 15, but for now just try to understand it at its face value: each n_k is just a natural number, so for any particular k we could let $i = n_k$. Then for that k, $x_{n_k} = x_i$, which is a perfectly respectable element of the set A.

This gives us another bijection

$$g : \{n_1, n_2, n_3, \ldots\} \to E, g : n_k \mapsto x_{n_k}, \forall k \in \mathbb{N}.$$

Notice that

$$(g \circ f)(k) = g(f(k)) = g(n_k) = x_{n_k},$$

so $g \circ f : \mathbb{N} \mapsto E$. Both f and g are bijections, so $g \circ f$ is a bijection (by the proof of Property 3 in Theorem 8.4). Thus $\mathbb{N} \sim E$, so E is countable.

To prove the second statement of the theorem, notice that the contrapositive of the first reads: "If E is not at most countable, then A is not at most countable." By definition, "not at most countable" just means uncountable, so if E is uncountable, then so is A. □

Corollary 8.12. *($\mathbb{R}$ Is Uncountable)*
The set of real numbers $\mathbb{R}$ is uncountable.

In Chapter 13, we will see a different proof of this theorem. For now, though, we just need to apply Theorem 8.11 to Example 8.10.

Proof. By Example 8.10, the open interval $(0, 1)$ is uncountable. But that open interval is a subset of the real numbers, so by Theorem 8.11, $\mathbb{R}$ is also uncountable. □

We'll use countable sets pretty often, so it will be useful to establish some facts about them while they're fresh in your mind (before you get distracted by all the awesome math in the next few chapters).

Theorem 8.13. *(Countable Union of Countable Sets)*
For every $n \in \mathbb{N}$, let E_n be a countable set. If $S = \bigcup_{n=1}^{\infty} E_n$, then S is also countable.

Proof. Since E_n is countable for any $n \in \mathbb{N}$, we can write it as $E_n = \{x_n^1, x_n^2, x_n^3, \ldots\}$. We'll also take advantage of the fact that we can write the sets in the union in order: $E_1, E_2, E_3, \ldots$

These facts give us a nice way to count the elements of S. Start with the first element of the first set x_1^1; then the second element of the first set x_1^2 followed by the first element of the second set x_2^1; then the third element of the first set x_1^3 followed by the second element of the second set x_2^2 and the first element of the third set x_3^1; and so on.

If we write the elements of S on a grid, in which the nth row is a list of the elements in E_n, this ordering has a nice visual representation:

$$E_1 = x_1^1 \quad x_1^2 \quad x_1^3 \quad \ldots$$

$$E_2 = x_2^1 \quad x_2^2 \quad x_2^3 \quad \ldots$$

$$E_3 = x_3^1 \quad x_3^2 \quad x_3^3 \quad \ldots$$

$$\vdots$$

Notice that this diagonal ordering is the same as counting in the following way: first, count all the elements x_i^j of S such that $i + j = 2$, then count all elements such that $i + j = 3$, then $i + j = 4$, and so on. (Within each set of all x_i^j such that $i + j =$ *some number*, we start with the elements with the smallest i, then the next smallest i, and so on.)

We now have a bijection $f \colon \mathbb{N} \to S$, so S is countable, right?

$$\begin{array}{cccccccccc} 1 & 2 & 3 & 4 & 5 & 6 & 7 & 8 & 9 & 10 & \ldots \\ \downarrow & \downarrow & \downarrow & \downarrow & \downarrow & \downarrow & \downarrow & \downarrow & \downarrow & \downarrow & \\ x_1^1 & x_1^2 & x_2^1 & x_1^3 & x_2^2 & x_3^1 & x_1^4 & x_2^3 & x_3^2 & x_4^1 & \ldots \end{array}$$

Wrong! What if, say, $x_1^2 = x_3^1$? Then f maps two different natural numbers, 2 and 6, to the same element of S, so f is *not* injective. Instead, let's have the function g be the same as f, except that it skips all duplicates in S—so if $g(n)$ would $= g(m)$ for some $m < n$, just don't even define g for that n. Let T be the subset of $\mathbb{N}$ where g is defined. Now we have $g \colon T \to S$, and g is indeed a bijection.

By Theorem 8.11, any subset of $\mathbb{N}$ is at most countable, so T is at most countable. Because g is a bijection, $S \sim T$. Then by Theorem 8.7, S is also at most countable. Since S cannot be finite (since it is the union of the *infinite* sets E_i), S must be countable. □

Note that in the previous theorem's proof, we used the fact that the number of unions is countable; in fact, the result is false for an uncountable union of countable sets (for example, just take $S = \bigcup_{\alpha \in \mathbb{R}} \{\alpha\}$, so $S = \mathbb{R}$ is uncountable). The following corollary will show that *any* countable union of countable sets is also countable (as opposed to the previous theorem, which only showed it for a union indexed over the natural numbers).

Corollary 8.14. *(Indexed Countable Union of Countable Sets)*
Let A be an at most countable set, and for each $\alpha \in A$, let E_α be an at most countable set. Then $T = \bigcup_{\alpha \in A} E_\alpha$ is also at most countable.

Proof. Let's start with the case in which A is countably infinite, and every E_α is countably infinite. Then A and the natural numbers can be put in one-to-one correspondence (remember Definition 8.3—this is just another way of saying $A \sim \mathbb{N}$), so to each $\alpha \in A$ there corresponds an $n \in \mathbb{N}$. Then we might as well index the sets in the union by their corresponding indexes in $\mathbb{N}$, so

$$T = \bigcup_{\alpha \in A} E_\alpha = \bigcup_{n \in \mathbb{N}} E_n = \bigcup_{n=1}^{\infty} E_n,$$

which puts T in the same form as S from Theorem 8.13, so T is countable.

Now we can look at the case in which A and/or any of the sets E_α are finite. We can just tag on elements to E_α to make it infinite: for each $\alpha \in A$, if E_α is finite, let $E'_\alpha = E_\alpha \cup \{1, 2, 3, \ldots\}$; if E_α is infinite, just let $E'_\alpha = E_\alpha$. Then each $E_\alpha \subset E'_\alpha$. To show that each E'_α is countably infinite, we can apply Theorem 8.13. Let the set $F_1 = E_\alpha$, and let $F_2 = \{1\}, F_3 = \{2\}, F_4 = \{3\}, \ldots$. So $\bigcup_{n=1}^{\infty} F_n = E'_\alpha$ is countable. Now we have:

$$T = \bigcup_{\alpha \in A} E_\alpha \subset \bigcup_{\alpha \in A} E'_\alpha,$$

and we can just apply the same logic in the previous case to get $\bigcup_{\alpha \in A} E'_\alpha = \bigcup_{n=1}^{\infty} E'_n$.

Thus whether A is finite or countably infinite, T is a subset of something in the same form as S from Theorem 8.13, so by Theorem 8.11, T is at most countable. □

Theorem 8.15. *(Tuples of Countable Sets)*
Let A be a countable set, and let A^n be the set of all n-tuples of A. Then A^n is countable.

The notation A^n means the same thing it does for $\mathbb{R}^n$: every element $\mathbf{a} \in A^n$ can be written $\mathbf{a} = (a_1, a_2, \ldots, a_n)$, where $a_i \in A$ for every i between 1 and n.

Proof. Since the number of cases we want to prove—which is the number of possible dimensions A^n can have—is countably infinite (since $n \in \mathbb{N}$), we can prove this by induction. Fill in the blanks in Box 8.1.

BOX 8.1

PROVING THEOREM 8.15 BY INDUCTION

Base Case. Assume $n = 1$. Then A^1 is countable because
_____.

Inductive Step. By the inductive hypothesis, we assume that _____ is countable. Every element of A^n can be written as $\mathbf{a} = (a_1, a_2, \ldots, a_n)$ where $a_1, a_2, \ldots, a_n \in A$, or equivalently as $\mathbf{a} = (\mathbf{b}, a_n)$ where $\mathbf{b} = (a_1, a_2, \ldots, a_{n-1})$.
If we fix $\mathbf{b}$, then we can make a bijection

$$f: \underline{\hspace{3cm}}, \quad f: \mathbf{a} \mapsto (\mathbf{b}, a_n).$$

Countability • 77

So for every $a^{n-1} \in A^{n-1}$, we have $A \sim \{(\mathbf{b}, a_n) \mid a_n \in A\}$. By Theorem _____, since A is countable, so is _____. We can write A^n as

$$A^n = \bigcup_{\mathbf{b} \in A^{n-1}} \{(\mathbf{b}, a_n) \mid a_n \in A\},$$

which by the inductive hypothesis is a _____ union of countable sets, so by Corollary _____, A^n is at most countable. Thus the fact that A is infinite implies A^n is _____.

□

Theorem 8.16. *($\mathbb{Q}$ Is Countable)*
The set of all rational numbers is countable.

Proof. Since every rational number can be written $\frac{a}{b}$, where $a, b \in \mathbb{Z}$, we can make a surjective function

$$f : \mathbb{Z} \times \mathbb{N} \to \mathbb{Q}, f : (a, b) \mapsto \frac{a}{b}.$$

Here $\mathbb{Z} \times \mathbb{N}$ is the set of doubles of the form $(a \in \mathbb{Z}, b \in \mathbb{N})$. (We need to do this—as opposed to using the more simple $\mathbb{Z}^2$—to avoid the possibility of $b = 0$ making an invalid fraction.)

In this case once again, though, f is not one-to-one—since, for example, both $(1, 2) \in \mathbb{Z} \times \mathbb{N}$ and $(3, 6) \in \mathbb{Z} \times \mathbb{N}$ map to the same rational number.

How do we get around this? As in the proof of Theorem 8.11, we'll define a "bijective version" g of f that acts on a subset of $\mathbb{Z} \times \mathbb{N}$. To make repetitive fractions unique, we'll use *gcd* function, which means *greatest common divisor*. Let

$$Z_n = \{k \in \mathbb{Z} \mid gcd(k, n) = 1\}.$$

So $Z_1 = \{1\}$, Z_2 is the set of all odd integers, Z_3 is the set of all integers that are not divisible by 3, and so on.

Let $T_n = \{(z, n) \in \mathbb{Z} \times \mathbb{N} \mid z \in Z_n\}$. In other words, T_n is the set of all integers that are not divisible by n, paired with n. If we set $T = \bigcup_{n \in \mathbb{N}} T_n$, then T is the set of all integer/natural pairs that are not divisible by each other. So the function

$$g : T \to Q, (z, n) \mapsto \frac{z}{n}$$

is a bijection. Why? It is surjective because every fraction can be written as $\frac{z}{n}$, and it is injective because each such fraction is unique (since z is not divisible by n).

Since $\mathbb{Z}$ is countable by Example 8.8, Theorem 8.15 says that $\mathbb{Z}^2$ is also countable, so by Theorem 8.11, any subset of $\mathbb{Z}^2$ is countable. T is a subset of $\mathbb{Z} \times \mathbb{N}$, which in turn is a subset of $\mathbb{Z}^2$, so now we have a bijection between a countable subset of $\mathbb{Z}^2$ and $\mathbb{Q}$. Thus $\mathbb{Q}$ is countable. □

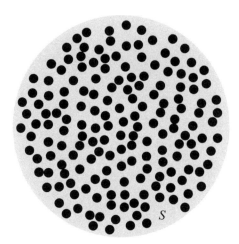

Figure 8.1. The set S is finite and thus at most countable.

Corollary 8.17. *(The Irrational Numbers Are Uncountable)*
The set of all irrational numbers is uncountable.

Proof. Let I be the set of all irrational numbers, so $\mathbb{R} = \mathbb{Q} \cup I$. We'll do a short proof by contradiction.

By Theorem 8.16, $\mathbb{Q}$ is countable. So if I were at most countable, then $\mathbb{R}$ would be a finite union of two at most countable sets, so by Corollary 8.14 $\mathbb{R}$ would be countable.

But by Corollary 8.12, $\mathbb{R}$ is not countable, so we have a contradiction. Thus I cannot be at most countable, so the set of irrational numbers must be uncountable. □

Because these notions of countability are pretty abstract, I'm afraid this wasn't a very visual-friendly chapter—so I thought you might like a picture.

Figure 8.1 is a good example of how a set that is technically "at most countable" may have a deceiving name; try to count all the dots in S.

On second thought, why don't you go do something productive instead.

Coming up next, we'll jump into the field of topology, starting with metric spaces and then moving on to plenty of other definitions. In a few chapters, we'll see the ways such definitions can help us learn even more about whether sets are countable.

CHAPTER 9

Topological Definitions

As we saw in Chapter 6, Euclidean spaces are not fields. Fields are all about *operations* (like addition and multiplication), whereas Euclidean spaces are all about *space*. We want to be able to work with spaces more abstract than $\mathbb{R}^k$, but that still have a way of relating any two elements to each other. In this chapter, we will describe *metric spaces* for that purpose.

To extract what we really need from topology—the theory of compactness and connectedness (in later chapters)—we first need to understand the properties that elements and subsets of metric spaces can have. This chapter is filled with many new definitions, which are listed here for convenience:

9.1 Metric Space	9.9 Limit Point	9.19 Interior Point
9.3 Bounded Set	9.13 Closed Set	9.21 Open Set
9.7 Neighborhood	9.16 Dense Set	9.24 Perfect Set

Definition 9.1. *(Metric Space)*
A **metric space** is a set X together with a function $d\colon X \times X \to \mathbb{R}$ such that for any elements p, q, and r of X, the following properties hold:

> Property 1. *(Distance)* $d(p,q) > 0$ if $p \neq q$, while $d(p,p) = 0$.
> Property 2. *(Symmetry)* $d(p,q) = d(q,p)$.
> Property 3. *(Triangle Inequality)* $d(p,q) \leq d(p,r) + d(r,q)$.

The elements of X are called **points**, and d is called the **distance function** or **metric**.

Example 9.2. *(Metric Spaces)*
We can use what we learned in Chapter 6 to verify that the set $\mathbb{R}^k$ with the distance function $d(\mathbf{p}, \mathbf{q}) = |\mathbf{p} - \mathbf{q}|$ is a metric space:

> Property 1. $d(\mathbf{p}, \mathbf{q}) = |\mathbf{p} - \mathbf{q}|$ is a real number, by Definition 6.10. If $\mathbf{p} \neq \mathbf{q}$, then $|\mathbf{p} - \mathbf{q}| > 0$ by Property 1 of Theorem 6.11, and

$$d(p,p) = |\mathbf{p} - \mathbf{p}| = |\mathbf{0}| = 0.$$

Property 2. We have

$$\begin{aligned}
d(p,q) &= |\mathbf{p} - \mathbf{q}| \\
&= |(-1)(\mathbf{q} - \mathbf{p})| \\
&= |-1|\,|\mathbf{q} - \mathbf{p}| \quad \text{(by Property 2 of Theorem 6.11)} \\
&= d(q, p).
\end{aligned}$$

Property 3. We apply the triangle inequality from Property 5 of Theorem 6.11 to see that

$$d(p, r) = |\mathbf{p} - \mathbf{q}| \leq |\mathbf{p} - \mathbf{r}| + |\mathbf{r} - \mathbf{q}| = d(p, r) + d(r, q).$$

When we set $k = 1$, we see that $\mathbb{R}$ can be a metric space. In general, when we say "the metric space $\mathbb{R}$," we really mean "the ordered field $\mathbb{R}$ together with the distance function $d(p, q) = |p - q|$."

Any subset Y of a metric space X with the same distance function d is also a metric space. Why? Well, $p, q \in Y$ implies $p, q \in X$, which implies that $d(p, q)$ satisfies the three properties of a metric.

Definition 9.3. *(Bounded Set)*
*A subset E of a metric space X is **bounded** if there is some point q in X such that the distance between q and any point of E is less than some fixed, finite real number M.*
In symbols, $E \subset X$ is bounded if:

$$\exists q \in X, \exists M \in \mathbb{R} \text{ such that } \forall p \in E, d(p, q) \leq M.$$

*A set that is not bounded is called **unbounded**.*

Distinction. This meaning of *bounded* is different from the one used in Definition 4.5. Being bounded above or below is a property of subsets of ordered fields, whereas being bounded in the sense of the current definition is a property of subsets of metric spaces.

Of course, any ordered field is also a metric space if we use the distance function $d(p, q) = |p - q|$. In $\mathbb{R}$—which is an ordered field and a metric space—for example, the set $(\infty, 3]$ is bounded above (by anything ≥ 3), but it is *not* bounded in the sense of the current definition, as we will see in the following example. In Theorem 9.6, we will give a concrete way of relating the two different definitions.

Example 9.4. *(Bounded Sets)*
The set $[-3, 3]$ in the metric space $\mathbb{R}$—and similarly the set $[-3, 3] \cap \mathbb{Q}$ in the metric space $\mathbb{Q}$—is bounded by the point $q = 3$ and the number $M = 6$, since no point in $[-3, 3]$ is a distance of more than 6 away from 3. We could also say it is bounded by the point $q = 3$ and the number $M = 100$, but it is usually more helpful to choose M as small as possible. $[-3, 3]$ is also bounded by the point $q = -3$ with the same $M = 6$, by the point $q = 0$ with $M = 3$, and many other options.

As you might be able to tell, boundedness is just a question of whether the set stretches on to infinity. The set $[-3, 3]$ is bounded because no number in that set is greater than 3, *and* no number in the set is less than -3. However, $(-\infty, 3]$ is *not* bounded, since there is no number $q \in \mathbb{R}$ that bounds it: for any M we choose, we will always be able to find some $p \in (-\infty, 3]$ such that $d(p, q) > M$. Basically, because $(-\infty, 3]$ goes on to infinity, there will always be a point that is far away from q.

The set $(-3, 3)$ in the metric space $\mathbb{R}$—and similarly the set $(-3, 3) \cap \mathbb{Q}$ in the metric space $\mathbb{Q}$—is bounded by the point $q = 3$ and the number $M = 6$. Remember: q must be in X (the metric space), but it need not be in E (the subset).

The set $\mathbb{Q}$ in the metric space $\mathbb{R}$, on the other hand, is unbounded, since for any point q of $\mathbb{R}$ with a distance of M away from some rational number, we can always find another rational number that is further than M away from q.

Theorem 9.5. *(Union of Bounded Sets)*
For any collection of subsets $\{A_i\}$ of a metric space X, if A_i is bounded for each i between 1 and n, then the finite union $\bigcup_{i=1}^{n} A_i$ is also bounded.

Proof. We can use the finiteness of the union to show that for any $q \in X$, $d(p, q)$ is less than the distance of the furthest away p_i, which is bounded by M_i.

For each A_i, there is a point $q_i \in X$ and a number $M_i \in \mathbb{R}$ such that $d(p_i, q_i) \leq M_i$ for each $p_i \in A_i$. For any point p of $\bigcup_{i=1}^{n} A_i$, we know $p \in A_i$ for some i between 1 and n, so

$$d(p, q_1) \leq d(p, q_i) + d(q_i, q_1)$$
$$\leq M_i + d(q_1, q_i)$$
$$\leq \max\{M_1, M_2, \ldots, M_n\} + \max\{d(q_1, q_1), d(q_1, q_2), \ldots, d(q_1, q_n)\}.$$

If we let

$$q = q_1, \text{ and } M = \max_{1 \leq i \leq n} M_i + \max_{1 \leq i \leq n} d(q_1, q_i),$$

then we have $d(p, q) \leq M$ for every $p \in \bigcup_{i=1}^{n} A_i$. □

Notice that this proof would not work if we had an *infinite* union $\bigcup_{i=1}^{\infty} A_i$, since we cannot necessarily take the maximum of an infinite set like $\{M_1, M_2, M_3, \ldots\}$.

Theorem 9.6. *(Bounded $\iff$ Bounded Above and Below)*
For any subset E of an ordered field F, E is bounded if and only if it is bounded above and below.

It is understood from the statement of this theorem that the ordered field F is also a metric space, with the distance function $d(p, q) = |p - q|$.

Proof. If E is bounded, there exists $q \in F$ and $M \in \mathbb{R}$ such that for every $p \in E$, we have

$$|p - q| \leq M \implies -M \leq p - q \leq M$$
$$\implies q - M \leq p \leq q + M.$$

Thus $q - M$ is a lower bound of E, and $q + M$ is an upper bound of E.

If E is bounded above and below, then there exist $\alpha, \beta \in F$ such that for every $p \in E$, we have $\beta \leq p \leq \alpha$. Choose

$$M \geq \max\{|\alpha|, |\beta|\},$$

so that $M \geq |\alpha| \geq \alpha$, and $-M \leq -|\beta| \leq \beta$. Because F is an ordered field, it has the additive identity element 0. Let $q = 0$, so $q - M \leq p \leq q + M$, so $|p - q| \leq M$. □

Definition 9.7. *(Neighborhood)*
A **neighborhood** $N_r(p)$ of **radius** $r > 0$ around a point p in a metric space X is the set of all points in X whose distance from p is less than r.
In symbols:

$$N_r(p) = \{q \in X \mid d(p, q) < r\}.$$

Example 9.8. (Neighborhoods)
In the metric space $\mathbb{R}$, the set $(-3, 3)$ is a neighborhood of radius 3 around the point 0. The set $(99, 100)$ is a neighborhood of radius 0.5 around the point 99.5. The sets $[-3, 3]$, $[-3, 3)$, and $(-3, 3]$ are not neighborhoods, since the points -3 and 3 are a distance of 3 (*equal* to the radius) away from 0.

In $\mathbb{R}^2$, the interior of any circle is a neighborhood around its center. The interior of any k-dimensional sphere is a neighborhood around its center in $\mathbb{R}^k$. (These sets are known as *open balls*.)

Note that all neighborhoods are bounded by the number $M = r$, since no point in the neighborhood can be further than r away from the center.

Definition 9.9. *(Limit Point)*
A point p is a **limit point** of a subset E of a metric space X if every neighborhood of p contains at least one point of E (other than possibly p itself).
In symbols, p is a limit point of $E \subset X$ if:

$$\forall r > 0,\ N_r(p) \cap E \neq \{p\} \text{ and } \neq \emptyset.$$

Limit points are also called **cluster points** *or* **accumulation points**. *Every non–limit point that is in E is called an* **isolated point** *of E.*

Example 9.10. (Limit Points)
Every point in the sets $[-3, 3]$ and $(-3, 3)$ is a limit point of that set. Why? By the density property of $\mathbb{Q}$ in $\mathbb{R}$, for every $p \in \mathbb{R}$ and every $r > 0$, we can find a point q such that $p - r < q < p + r$. Thus q is in $N_r(p)$. Now we just need a way to guarantee that $q \in (-3, 3)$. Let's use the density property to instead find a q_1 with $\max(p - r, -3) < q_1 < \min(p + r, 3)$. Then $q_1 \in N_r(p)$ since

$$-3 \leq \max(p - r, -3) < q_1 < \min(p + r, 3) \leq 3,$$

and also $q_1 \in (-3, 3)$ since $-3 < q_1 < 3$.

Notice that by the same logic, -3 and 3 are limit points of the open interval $(-3, 3)$ (even though they are not elements of that interval).

Topological Definitions • 83

By definition, every neighborhood of a limit point $p \in E$ contains at least one other point of E. It turns out that we can actually say more: every neighborhood of p contains an *infinite* number of points of E. That seems like a pretty big jump, doesn't it? We go from knowing there is one point of E to knowing there are infinitely many points of E. But the key here is that *every* neighborhood of p contains a point of E; there are infinitely many neighborhoods, so each one should have infinitely many points.

Theorem 9.11. *(Infinite Neighborhoods of Limit Points)*
If p is a limit point of a subset E of a metric space X, then for any $r > 0$, $N_r(p)$ contains infinitely many points of E.

Proof. Let's do a proof by contrapositive: Instead of showing **A** $\implies$ **B**, we will show $\neg$**B** $\implies$ $\neg$**A**. In this case, **B** is "every neighborhood of p contains *infinitely* many points of E", so $\neg$**B** is "*some* neighborhood of p contains *finitely* many points of E." The basic idea is that if the number of points is finite, we can choose the one with minimum distance, and the neighborhood of radius less than that distance will only contain p and nothing else.

Fill in the blanks in Box 9.1.

BOX 9.1

PROVING THE CONTRAPOSITIVE OF THEOREM 9.11

Assume $\exists r > 0$ such that $N_r(p) \cap E$ contains a _____ number of points of E. Then $Q = (N_r(p) \cap E) \setminus \{p\}$ (the neighborhood without the point p) also contains a finite number of points. If Q is empty, then we are done (we can skip to the last line of this proof). Otherwise, we can label all the points of Q as $\{q_1, q_2, \ldots, q_n\}$, and let the set

$$D = \{d(p, q_1), d(p, q_2), \ldots, d(p, q_n)\}.$$

Since Q is finite, so is D, and we can take the minimum.
Let $h = \frac{1}{2}$ _____. Every element of D is positive, so h is also _____. Now $N_h(p) \cap Q =$ _____.
Since $h \leq r$, we know $N_h(p) \subset N_r(p)$, so

$$N_h(p) \cap E = \big(N_h(p) \cap N_r(p)\big) \cap E = N_h(p) \cap \big(N_r(p) \cap E\big) = N_h(p) \cap Q = \underline{}.$$

We have found a neighborhood of p that contains no points of E (except for possibly p itself), so p cannot be a _____ of E.

□

Corollary 9.12. *(Finite Sets Have No Limit Points)*
A finite subset of a metric space has no limit points.

Proof. If a finite $E \subset X$ had a limit point p, then by Theorem 9.11, for any $r > 0$, $N_r(p)$ would contain infinitely many points of E. Then E must have an infinite number of points, which is a contradiction. □

Definition 9.13. *(Closed Set)*
*A subset of a metric space is **closed** if it contains all of its limit points.*
   In symbols, $E \subset X$ is closed if:

$$\{p \in X \mid p \text{ is a limit point of } E\} \subset E.$$

Distinction. This meaning of *closed* is different from the one used in Definition 5.1. A field can be closed under addition and/or multiplication; only a subset of a metric space can be closed because it contains all of its limit points.

Example 9.14. *(Closed Sets)*
The closed interval $[-3, 3]$ is closed. (Now you finally know why we call it a "closed interval"!) There aren't any points p outside of $[-3, 3]$ that are limit points of $[-3, 3]$, because with

$$r = \frac{\min\{|-3 - p|, |3 - p|\}}{2},$$

$N_r(p)$ doesn't contain any points of $[-3, 3]$.
   However, $(-3, 3)$ is not closed. By Example 9.10, we know that the points -3 and 3 are limit points of the open interval $(-3, 3)$. But -3 and 3 do not belong to that set.

Corollary 9.15. *(Finite Subsets Are Closed)*
A finite subset of a metric space is closed.

Proof. By Corollary 9.12, a finite set $E \subset X$ has no limit points. Thus E contains all of its limit points, so it is closed. □

Definition 9.16. *(Dense Set)*
*A subset E of a metric space X is **dense** in X if every point of X is a point of E and/or a limit point of E.*
   In symbols, $E \subset X$ is dense in X if:

$$\forall x \in X, \, x \in E \text{ and/or } x \text{ is a limit point of } E.$$

Note that density is a *relative* property. A set cannot be just plain "dense"; it only makes sense to talk about a set being dense *in* some metric space.

Example 9.17. *(Dense Sets)*
Notice that every metric space is dense in itself.
   If a set E in a metric space X is closed, then E contains all of its limit points, so saying E is dense in X is the same as saying that every point of X is a point of E, meaning $X \subset E$. But $E \subset X$, so we have the following rule: for any closed subset E of a metric space X, E is dense in X if and only if $E = X$.

Distinction. This meaning of *dense* is different from the one used in Theorem 5.6. For example, according to this definition, any set E is dense in itself (which is not true of $\mathbb{N}$ according to Theorem 5.6's definition).

However, it turns out that $\mathbb{Q}$ is dense in $\mathbb{R}$, in both the old and the new meaning of the term.

Theorem 9.18. *($\mathbb{Q}$ Is Dense in $\mathbb{R}$)*
The set of rational numbers $\mathbb{Q}$ is dense in the metric space $\mathbb{R}$.

Proof. We want to show that every real number is either a rational number, or a limit point of the set of rational numbers (or both). Well, every real number that is not rational is irrational, so all we need to show is that every irrational number is a limit point of $\mathbb{Q}$. In other words, if we let I be the set of all irrational numbers, we must show that

$$p \in I \implies \forall r > 0, N_r(p) \cap \mathbb{Q} \neq \{p\} \text{ and } \neq \emptyset.$$

Remember that on the real line, a neighborhood is just an open interval. Meaning for any $p \in I$, $N_r(p) = (p - r, p + r)$, so we need there to be some $q \in \mathbb{Q}$ such that $p - r < q < p + r$. Notice that since $p - r$ and $p + r$ are real numbers, Theorem 5.6 guarantees that such a rational number q exists. $\square$

The following definition will help us introduce *open sets*, which are in a way the opposite of closed sets.

Definition 9.19. *(Interior Point)*
*A point p is an **interior point** of a subset E of a metric space X if some neighborhood of p is contained in E.*

In symbols, p is an interior point of $E \subset X$ if:

$$\exists r > 0 \text{ such that } N_r(p) \subset E.$$

Example 9.20. (Interior Points)
Every point of $(-3, 3)$ is an interior point of that set. Why? For any p in $(-3, 3)$, let

$$r = \min\{|-3 - p|, |3 - p|\}.$$

Then $N_r(p) \subset (-3, 3)$.
Every point of $[-3, 3]$ except for -3 and 3 is an interior point of that set. Why are those two numbers not interior points? For every $r > 0$, $N_r(-3)$ contains a point less than -3, so no neighborhood of -3 is a subset of $(-3, 3)$. Similarly, every neighborhood of 3 contains a point greater than 3.

Definition 9.21. *(Open Set)*
*A subset of a metric space is **open** if all of its points are interior points of that set.*
In symbols, $E \subset X$ is open if:

$$\forall p \in E, \exists r > 0 \text{ such that } N_r(p) \subset E.$$

Example 9.22. (Open Sets)
By the previous example, $(-3, 3)$ is open, and $[-3, 3]$ is not open.

We will next prove that all neighborhoods are open. Note that just because $[-3, 3]$ is closed but not open while $(-3, 3)$ is open but not closed, does not mean that *open* ⟹ *not closed*, or vice versa. We will see examples of sets that can be neither open nor closed, and sets that are both open and closed.

Theorem 9.23. *(Neighborhoods Are Open)*
Every neighborhood $N_r(p)$ in a metric space X is an open set.

Proof. First, we'll break the proof down step by step by using the definitions to figure out what the crux of the problem is. Second, we'll take what we have and write it up in a nice, linear fashion.

Step 1. Let's narrow down what we need to prove by applying definitions.
↪ We want to prove that every neighborhood is an open set.
  ↪ Take a neighborhood $E = N_r(p)$, and show that every $q \in E$ is an interior point of E.
    ↪ Show that $\forall q \in E, \exists k > 0$ such that the neighborhood $F = N_k(q)$ is contained in E.
      ↪ Show that $\forall s \in F, s \in E$.
        ↪ Show that $\forall s \in F, d(p, s) < r$.

If we can find a radius k for the neighborhood F that makes this last statement true, we are done. (We can choose any $k > 0$ that works, because for q to be an interior point, we just need *one* neighborhood that fits inside E.)

To get started, let's consider the facts available to us that we haven't yet used:

1. X is a metric space ⟹ we have the triangle inequality.
2. $q \in E$, so $d(p, q) < r$.

Thus $d(p, s) \leq d(p, q) + d(q, s) < r + k$.

k is an arbitrary real number, but it must be > 0. So we aren't quite finished—just because $d(p, s) < r + $ anything that is bigger than 0, doesn't mean $d(p, s) < r$.

Look at the second fact again. If $d(p, q) < r$, there must be an $h > 0$ such that $d(p, q) + h = r$. Then

$$d(p, s) \leq r - h + d(q, s) < r - h + k.$$

Now look at Figure 9.1, and it's clear! Just let $k = h$, so that $d(p, s) < r$.

Step 2. Yes, we just proved the theorem, but notice that if we make use of that critical step (letting $k = h$) at the beginning, we can write everything more cleanly. Here's the formal proof.

Let $E = N_r(p)$, so for any $q \in E$, $d(p, q) < r$ implies that there is some $h > 0$ such that $d(p, q) + h = r$. Let $F = N_h(q)$, so for any $s \in F$, we have $d(q, s) < h$. X is a metric space, so by the triangle inequality, we have

$$d(p, s) \leq d(p, q) + d(q, s) = r - h + d(q, s) < (r - h) + h = r,$$

so $s \in N_r(p) = E$. But s was arbitrary, so we can say that *every* point in $N_h(q)$ is in E, so q is an interior point of E. But q was arbitrary, too, so we can say that *every* point in E is an interior point of E, and thus E is an open set.

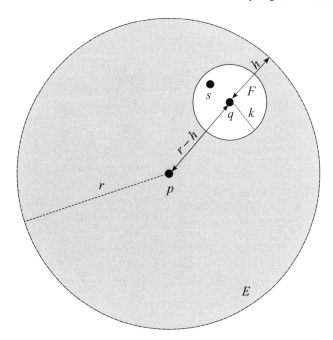

Figure 9.1. The point q is an interior point of $N_r(p)$.

If we want, we could shorten the proof even more by putting almost everything in symbols:

$$\forall q \in E, d(p,q) < r \implies \exists h > 0 \text{ such that } d(p,q) = r - h,$$
$$\text{and } \forall s \in N_h(q), d(q,s) < h$$
$$\implies d(p,s) \leq d(p,q) + d(q,s) < r - h + h = r$$
$$\implies s \in E$$
$$\implies N_h(q) \subset E$$
$$\implies E \text{ open.}$$

$\square$

Definition 9.24. *(Perfect Set)*
*A subset of a metric space is **perfect** if it is closed and if all of its points are limit points.*

In other words, a subset of a metric space is perfect if its limit points are exactly all of its points.

Example 9.25. (Perfect Sets)
Of the examples we have seen so far, only closed intervals such as $[-3, 3]$ are perfect. A set such as $[-3, 3] \cup \{100\}$ is closed but not perfect, because 100 is not a limit point.

Example 9.26. (Properties of Different Sets)
For each subset E of the metric space X, we'll look at whether it is bounded and then at its limit points and interior points to determine whether it is closed, open, and/or perfect.

1. $E = \emptyset$.
   Bounded? Yes, by any point $q \in X$ and any number M.
   Limit points? None, since E has no points.
   Interior points? None, since E has no points.
   Closed? Yes. E has no limit points, so it contains all of them.
   Open? Yes. E has no points, so all of its points are interior points.
   Perfect? Yes. E is closed, and it has no points, so all of its points are limit points.
2. $E = \{p\}$.
   Bounded? Yes, by any point $q \in X$ and the number $M = d(p, q)$. Because for any $q \in X$, the distance between q and any point of E is actually $d(p, q)$, and $d(p, q) \leq d(p, q) = M$.
   Limit points? None. Each neighborhood around p only contains one element of E, which is p itself. For every other point $q \in X$, choose $0 < r < d(p, q)$, so $N_r(q) \cap E = \emptyset$.
   Interior points? It completely depends on what our metric space X is. If X contains a finite number of points $\{q_1, q_2, \ldots, q_n\}$, then we set

   $$r = \frac{1}{2}\min\{q_1, q_2, \ldots, q_n\},$$

   and $N_r(p)$ contains only p, so p is an interior point of E.
   If X contains an infinite number of points, then this trick does not work—we cannot find the point of X with the minimum distance to p, since it is impossible to take the minimum of an infinite set (only the infimum, which doesn't help in this example). Here there are two possible cases.
   Case 1. p is *not* an interior point of E: if, for example, $X = \mathbb{R}$ with the usual metric, every neighborhood around p contains non-E points of X.
   Case 2. p *is* an interior point of E: if, for example, $X \subset \mathbb{R}$ with $X = (-\infty, -1] \cup \{0\} \cup [1, \infty)$ and $p = 0$, then $N_{0.5}(p)$ contains only one point in X, namely, p.
   Closed? Yes. E has no limit points, so it contains all of them.
   Open? Again, it depends on the metric space. If p is an interior point of E, then E is open, since all of its points are interior points. Otherwise, E is not open.
   Perfect? No. E is closed, but its only point p is not a limit point.
   As this example demonstrates, we usually need to be careful to specify which metric space we are working in.
3. $E = [-3, 3]$, in the metric space $X = \mathbb{R}$.
   Bounded? Yes, by Example 9.4.
   Limit points? By Example 9.10, every point of E is a limit point.
   Interior points? By Example 9.20, every point of E except for -3 and 3 is an interior point.
   Closed? Yes. E contains all of its limit points.
   Open? No. There are points of E—namely, -3 and 3—that are not interior points, so E cannot be open.
   Perfect? Yes. E is closed and all of its points are limit points.
4. $E = (-3, 3)$, in the metric space $X = \mathbb{R}$.
   Bounded? Yes, by Example 9.4.

Limit points? By Example 9.10, every point of E is a limit point. Also, -3 and 3 are limit points of E.

Interior points? By Example 9.20, every point of E is an interior point.

Closed? No. There are two limit points of E—namely, -3 and 3—that are not contained in E, so E cannot be closed.

Open? Yes. Every point of E is an interior point.

Perfect? No. E is not closed.

5. $E = (-3, 3]$, in the metric space $X = \mathbb{R}$.

    Bounded? Yes, by Example 9.4.

    Limit points? By Example 9.10, every point of E is a limit point. Also, -3 is a limit point of E.

    Interior points? Every point of E is an interior point, except for 3.

    Closed? No. There is a limit point of E—namely, -3—that is not contained in E, so E cannot be closed.

    Open? No. E contains a point—namely, 3—which is not an interior point.

    Perfect? No. E is not closed.

Note that this last example is neither closed nor open.

In the next chapter, we'll explore openness and closedness in more depth by proving several theorems, defining the *closure* of a set, and learning about how a set can be open relative to one set but not another. (I know you can't wait to turn the page, but you should probably go see your friends first. They miss you.)

CHAPTER 10

Closed and Open Sets

Whereas the last chapter was mostly concerned with definitions, this chapter will present some important theorems that should help us better understand how to work with open and closed sets.

Theorem 10.1. *(Complement of Open Sets)*
A subset E of a metric space X is open if and only if its complement is closed.

If you think of open sets as having "dotted line" borders (i.e., not containing their boundaries) and of closed sets as having "solid line" borders, then the theorem should make sense. As in Figure 10.1, if a set has a dotted line border, then the points outside of it include the points of that border.

Proof. The proof can be very concise. If E^C is closed, then $x \in E$ implies x cannot be a limit point of E^C, so some neighborhood around x contains only points of E, so E is open. Conversely, if E is open, then for any limit point x of E^C, no neighborhood of x contains only points of E, so x cannot be in E, and thus E^C is closed.

In case you're still hazy on limit points and interior points, we'll expand this proof by filling in all the little steps.

To prove one direction of the theorem, assume E^C is closed. We want to show that E is open, meaning some neighborhood around every point in E is completely contained in E. Well, E^C contains all of its limit points, so for any $x \in E$, we have $x \notin E^C$, so x cannot be a point of E^C nor a limit point of E^C. Then the definition of a limit point is false for x, so there must exist some $r > 0$ such that $N_r(x) \cap E^C = \{x\}$ or $= \emptyset$. Because $x \notin E^C$, $N_r(x) \cap E^C$ cannot contain x, hence $N_r(x) \cap E^C$ must $= \emptyset$. Because $N_r(x)$ shares no points with E^C, it only contains points of E, so $N_r(x) \subset E$. Thus x is an interior point of E. And since x was arbitrary, we have that every point of E is an interior point, so E is an open set.

To prove the other direction, assume E is open. We want to show that E^C is closed, meaning E^C contains all of its limit points. If x is a limit point of E^C, then for any $r > 0$, $N_r(x) \cap E^C$ contains at least one point of E^C. Thus $N_r(x)$ does not contain only points of E, so every neighborhood of x is not entirely contained in E, so x cannot be an interior point of E. But every point of E must be an interior point, so x cannot be in E, so $x \in E^C$. Since x was an arbitrary limit point of E^C, we have that every limit point of E^C is inside E^C, so E^C is a closed set. □

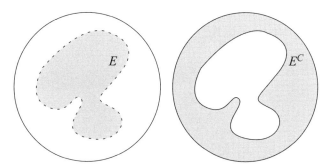

Figure 10.1. The open set E is the shaded area on the left (which excludes the dotted border), and the closed set E^C is the shaded area on the right (which includes the solid border).

Corollary 10.2. *(Complement of Closed Sets)*
A subset F of a metric space X is closed if and only if its complement is open.

Proof. If F^C is open, then by the previous theorem, the complement of F^C must be closed. But $\left(F^C\right)^C$ is the same as F, so F is closed. Similarly, if F is closed, then the set whose complement is F must be open, and this set is clearly F^C. □

The previous theorem and corollary give us a good way of relating closedness to openness, but don't let your guard down. Always remember that open is *not* the opposite of closed (since sets can be both open and closed, or neither—see Example 9.26).

Theorem 10.3. *(Infinite Open Unions and Closed Intersections)*
The union of any collection of open sets is also open.
   In symbols:

$$G_\alpha \text{ open } \forall \alpha \implies \bigcup_\alpha G_\alpha \text{ open.}$$

Similarly, the intersection of any collection of closed sets is also closed.
   In symbols:

$$F_\alpha \text{ closed } \forall \alpha \implies \bigcap_\alpha F_\alpha \text{ closed.}$$

Proof. Every point of $\bigcup_\alpha G_\alpha$ is an interior point of some G_α, so it has some neighborhood completely contained in the union. Every neighborhood of every limit point of $\bigcap_\alpha F_\alpha$ intersects each F_α, so it is a limit point of each F_α, so it is an element of the intersection. Let's formalize this argument.

For any $x \in \bigcup_\alpha G_\alpha$, we know $x \in G_\alpha$ for some α. Now G_α is open, so x is an interior point of G_α, so $\exists r > 0$ such that $N_r(x) \subset G_\alpha$. But $G_\alpha \subset \bigcup_\alpha G_\alpha$, so $N_r(x)$ is also contained in $\bigcup_\alpha G_\alpha$, so x is also an interior point of $\bigcup_\alpha G_\alpha$. Since x was arbitrary, we have that every point of $\bigcup_\alpha G_\alpha$ is an interior point, so it is an open set.

Let x be a limit point of $\bigcap_\alpha F_\alpha$, so for any $r > 0$,

$$N_r(x) \cap \left(\bigcap_\alpha F_\alpha\right) \neq \{x\} \text{ and } \neq \emptyset.$$

This means that for *every* α, $N_r(x) \cap F_\alpha \neq \{x\}$ and $\neq \emptyset$. (Why? If the intersection of the neighborhood and *all* the F_α's contains something, then the intersection of the neighborhood and any one F_α certainly contains something as well.) Since F_α is closed for every α, we see that $x \in F_\alpha$ for every α, so $x \in \bigcap_\alpha F_\alpha$. Since x was an arbitrary limit point, we have that every limit point of $\bigcap_\alpha F_\alpha$ is a point of $\bigcap_\alpha F_\alpha$, so it is a closed set.

Notice the similarity between the two arguments—we took an arbitrary point/limit point and just applied the definitions of open/closed and union/intersection. Say, though, you got stuck on proving the second part... You could have just applied Corollary 10.2 and De Morgan's law as follows: since F_α is closed, F_α^C is open, so by the proof of the first property, $\bigcup_\alpha \left(F_\alpha^C\right)$ is also open. We know by De Morgan's law that $\left(\bigcap_\alpha F_\alpha\right)^C = \bigcup_\alpha \left(F_\alpha^C\right)$, so $\bigcap_\alpha F_\alpha$ must be closed. □

Theorem 10.4. *(Finite Open Intersections and Closed Unions)*
The intersection of any finite collection of open sets is also open.
   In symbols:

$$G_i \text{ open for } 1 \leq i \leq n \implies \bigcap_{i=1}^n G_i \text{ open}.$$

Similarly, the union of any finite collection of closed sets is also closed.
   In symbols:

$$F_i \text{ closed for } 1 \leq i \leq n \implies \bigcup_{i=1}^n F_i \text{ closed}.$$

The difference in this theorem (from the previous theorem) is that the collection of sets must be finite.

As we saw in Example 3.14, if $A_n = (-\frac{1}{n}, \frac{1}{n})$ for any $n \in \mathbb{N}$, then $\bigcap_{n=1}^\infty A_n = \{0\}$. Each A_n is an open set in $\mathbb{R}$, but their *infinite* intersection is just a single point, which is not open in $\mathbb{R}$.

Similarly, we also saw in Example 3.14 that if $A_n = [0, 2 - \frac{1}{n}]$ for any $n \in \mathbb{N}$, then $\bigcup_{n=1}^\infty A_n = [0, 2)$. We never really proved this, though, so let's do it now. For any $0 \leq x < 2$, the Archimedean property provides an $n \in \mathbb{N}$ such that $n > \frac{1}{2-x}$, so that

$$2n - 1 > nx \implies x < 2 - \frac{1}{n}.$$

Thus any $x \in [0, 2)$ is in $[0, 2 - \frac{1}{n}]$ for some $n \in \mathbb{N}$, so that $x \in \bigcup_{n=1}^\infty [0, 2 - \frac{1}{n}]$. (The union does not, however, contain the number 2, since there is no $n \in \mathbb{N}$ such that $2 \in A_n$.) Each A_n is a closed set in $\mathbb{R}$, but their *infinite* union is a half-open interval, which is not closed in $\mathbb{R}$ (since 2 is a limit point of $[0, 2)$ but is not contained in it).

It's not that infinite intersections of open sets are *never* open; sometimes they are, and sometimes they aren't. For example, let $A_n = (-3, 3)$ for any $n \in \mathbb{N}$. Then each A_n is an open set in $\mathbb{R}$, and $\bigcap_{n=1}^\infty A_n = (-3, 3)$ is also open in $\mathbb{R}$. The same comment applies to infinite unions of closed sets.

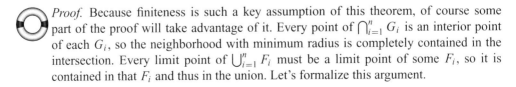

Proof. Because finiteness is such a key assumption of this theorem, of course some part of the proof will take advantage of it. Every point of $\bigcap_{i=1}^n G_i$ is an interior point of each G_i, so the neighborhood with minimum radius is completely contained in the intersection. Every limit point of $\bigcup_{i=1}^n F_i$ must be a limit point of some F_i, so it is contained in that F_i and thus in the union. Let's formalize this argument.

Take $x \in \bigcap_{i=1}^{n} G_i$. We want to show that x is an interior point of that intersection. For each i between 1 and n, we have $x \in G_i$, and since each G_i is open, there exists an $r_i > 0$ such that $N_{r_i}(x) \subset G_i$. Because n is finite, we can take the minimum such radius, so let $r = \min\{r_1, r_2, \ldots, r_n\}$. Then for each i between 1 and n,

$$N_r(x) \subset N_{r_i}(x) \subset G_i.$$

Since $N_r(x)$ is a subset of *every* G_i, we have $N_r(x) \subset \bigcap_{i=1}^{n} G_i$, and thus x is indeed an interior point of $\bigcap_{i=1}^{n} G_i$.

Let x be a limit point of $\bigcup_{i=1}^{n} F_i$. We want to show that x is an element of that union. Well, if x is a limit point of some F_i, then it is an element of that F_i, so $x \in \bigcup_{i=1}^{n} F_i$. We just need to show that x is a limit point of some F_i, which we'll do by contradiction. Assume x is not a limit point of any F_i, so for each i between 1 and n, there is some $r_i > 0$ such that $N_{r_i}(x) \cap F_i = \{x\}$ or $= \emptyset$. Since n is finite, we can take the minimum such radius, so let $r = \min\{r_1, r_2, \ldots, r_n\}$. Then for each i between 1 and n,

$$\left(N_r(x) \cap F_i\right) \subset \left(N_{r_i}(x) \cap F_i\right) = \{x\} \text{ or } = \emptyset.$$

Thus $N_r(x) \cap \left(\bigcup_{i=1}^{n} F_i\right) = \{x\}$ or $= \emptyset$, which contradicts the fact that x is a limit point of $\bigcup_{i=1}^{n} F_i$.

To prove the second statement, again we could have just applied Theorem 10.1 and De Morgan's law as in Box 10.1.

BOX 10.1

> PROVING THAT THE UNION OF FINITELY MANY CLOSED SETS IS CLOSED
>
> Since F_i is closed, F_i^C is _____, so by the proof of the first property, $\bigcap_{i=1}^{n} \left(F_i^C\right)$ is also _____. We know by De Morgan's law that $\bigcap_{i=1}^{n} \left(F_i^C\right) =$ _____, so because its complement is _____, $\bigcup_{i=1}^{n} F_i$ must be closed.

Definition 10.5. *(Closure)*
*The set of all limit points of a subset E of a metric space X is written E'. The **closure** of E, written $\overline{E}$, is the set $E \cup E'$.*

Example 10.6. *(Closures)*
If $E = (-3, 3)$, then by Example 9.10 we have $-3, 3 \in E'$. Also, every point of $(-3, 3)$ is a limit point, so $(-3, 3) \subset E'$. Thus $E' = \{-3, 3\} \cup (-3, 3) = [-3, 3]$, and $\overline{E} = (-3, 3) \cup [-3, 3]$, which is also the closed interval $[-3, 3]$.

Theorem 10.7. *(Any Closure Is Closed)*
For any subset E of a metric space X, $\overline{E}$ is closed.

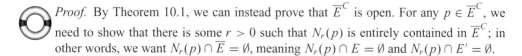

Proof. By Theorem 10.1, we can instead prove that $\overline{E}^C$ is open. For any $p \in \overline{E}^C$, we need to show that there is some $r > 0$ such that $N_r(p)$ is entirely contained in $\overline{E}^C$; in other words, we want $N_r(p) \cap \overline{E} = \emptyset$, meaning $N_r(p) \cap E = \emptyset$ and $N_r(p) \cap E' = \emptyset$.

94 • Chapter 10

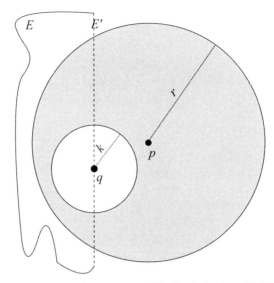

Figure 10.2. The set E has limit points represented by the dashed line E'. If $N_r(p) \cap E'$ contains a point q, then $N_r(p)$ must contain at least one point of E.

Well, $p \in \overline{E}^C$ implies $p \notin E$ and p is not a limit point of E, so there is some $r > 0$ such that $N_r(p) \cap E = \emptyset$.

Now all we need to show is that $N_r(p) \cap E' = \emptyset$. We'll do this by contradiction, so assume there is a q such that $q \in N_r(p) \cap E'$, as in Figure 10.2. This q has two properties: $d(p, q) < r$ (since q is in p's neighborhood), and q is a limit point of E (since $q \in E'$). Choose $k < r - d(p, q)$, so that $N_k(q) \subset N_r(p)$. Because q is a limit point of E, $N_k(q)$ must contain at least one point of E, so $N_r(p)$ contains at least one point of E, which contradicts what we just showed in the previous paragraph. □

Corollary 10.8. *(Equal to Closure $\iff$ Closed)*
For any subset E of a metric space X, $E = \overline{E}$ if and only if E is closed.

Proof. If $E = \overline{E}$, then since (by the previous theorem) $\overline{E}$ is closed, so is E. If E is closed, then $E' \subset E$ by the definition of closed, so $E = E \cup E' = \overline{E}$. □

Corollary 10.9. *(The Closure Is a Subset of Every Closed Superset)*
For any subset E of a metric space X, if $E \subset F$ for any closed set $F \subset X$, then $\overline{E} \subset F$.

This corollary demonstrates that the closure of a set E is the smallest possible closed set containing E (since every closed superset of E contains the closure $\overline{E}$). That statement should make intuitive sense, because if you want to build a closed set containing E, you cannot leave out any limit points of E.

Proof. If $E \subset F$, then the limit points of F must at least include the limit points of E, meaning $E' \subset F'$. If F is also closed, then $F' \subset F$, so $E' \subset F$, and thus $F \supset (E \cup E') = \overline{E}$. □

The next theorem helps relate limit points to suprema in $\mathbb{R}$, giving us a bridge between the theory of ordered sets and the theory of metric spaces (at least for real numbers). Remember that most metric spaces are not ordered fields (for example, $\mathbb{R}^2$).

Theorem 10.10. *(Real Suprema and Infima in Closures)*
For any nonempty subset E of $\mathbb{R}$ that is bounded above, $\sup E \in \overline{E}$. If E is bounded below, then $\inf E \in \overline{E}$.

Proof. Assume E is bounded above and let $y = \sup E$. If $y \in E$, then clearly $y \in \overline{E}$, so we only need to worry about the case where $y \notin E$. From the definition of supremum, we know that every element of E is $\leq y$, and that any number less than y is not an upper bound of E. Thus for every $h > 0$, there is some $x \in E$ such that $y - h < x < y$ (otherwise $y - h$ would be an upper bound of E). In other words, $x \in N_h(y) \cap E$. Since h was arbitrary, this shows that y is a limit point of E. Thus $y \in \overline{E}$.

The proof for infima is basically the same. Try copying the above argument by filling in the blanks in Box 10.2.

BOX 10.2

> **PROVING THEOREM 10.10 FOR INFIMA**
>
> Assume _____, and let $y = \inf E$. If $y \in E$, then $y \in$ _____. If not, then every element of E is _____ y, and any number greater than y is not a _____ of _____. Thus for every $h > 0$, there is some $x \in E$ such that $y < x <$ _____ (otherwise _____ would be a lower bound of _____). In other words, $x \in N_h(y) \cap E$. Since h was arbitrary, this shows that y is a _____ of E. Thus _____ $\in \overline{E}$.

Corollary 10.11. *(Real Bounded Closed Sets Contain sup and inf)*
For any nonempty subset E of $\mathbb{R}$ that is bounded above, if E is closed then $\sup E \in E$. If E is bounded below and closed, then $\inf E \in E$.

Note that the converse of this corollary ($\sup E \in E \implies E$ is closed) is not necessarily true. For example, the set $E = [-3, 0) \cup (0, 3]$ in $\mathbb{R}$ contains its infimum and and its supremum (-3 and 3, respectively), but is not closed, since 0 is a limit point.

Proof. By the previous theorem, we know $\sup E \in \overline{E}$ (if E is bounded above) and $\inf E \in \overline{E}$ (if E is bounded below). By Corollary 10.8, if E is closed, then $E = \overline{E}$, so indeed $\sup E \in E$ (if E is bounded above), and $\inf E \in E$ (if E is bounded below). □

Now is a good time to point out that most of our topological characteristics for sets completely depend on which metric space the set is considered to be in. For example, as we saw in Example 9.26, the open interval $(-3, 3)$ is open in $\mathbb{R}$—but it is *not* open in $\mathbb{R}^2$, since a neighborhood in $\mathbb{R}^2$ is a disk, and any disk contains points with 2-dimensional coordinates, which are not part of the 1-dimensional open interval $(-3, 3)$. In order to make a rule to solve this possible ambiguity, we should first formally define relative openness.

Definition 10.12. *(Relative Openness)*
A set E is **open relative to** Y if, for each $p \in E$, there is an $r > 0$ such that

$$q \in Y \text{ and } d(p, q) < r \implies q \in E.$$

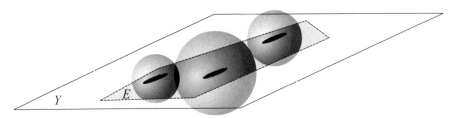

Figure 10.3. The set E is a pentagon in the 2-dimensional metric space Y. Each point of E has a neighborhood that is completely contained in E, so we want to extend those neighborhoods into spheres in the 3-dimensional metric space X.

This is basically the same as Definition 9.21. The only difference is that for every $p \in E$, instead of requiring $N_r(p) \subset E$ for some $r > 0$, we require $N_r(p) \cap Y \subset E$ for some $r > 0$.

Example 10.13. (Relative Openness)
In the previous example, we saw that $E = (-3, 3)$ is open relative to $Y = \mathbb{R}$, but not open relative to $Y = \mathbb{R}^2$.

Of course, for any subset E of a metric space X, saying "E is open relative to X" is the same as saying "E is open," since X is the entire metric space.

Theorem 10.14. (Relative Openness)
For any metric spaces $Y \subset X$, a subset E of Y is open relative to Y if and only if $E = Y \cap G$ for some open subset G of X.

In other words, every relatively open subset E of Y can be represented as the Y-part of an open subset G of X.

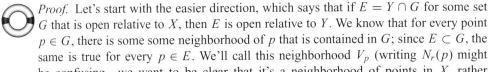

Proof. Let's start with the easier direction, which says that if $E = Y \cap G$ for some set G that is open relative to X, then E is open relative to Y. We know that for every point $p \in G$, there is some some neighborhood of p that is contained in G; since $E \subset G$, the same is true for every $p \in E$. We'll call this neighborhood V_p (writing $N_r(p)$ might be confusing—we want to be clear that it's a neighborhood of points in X, rather than a neighborhood of only points in Y). We just intersect Y with both sides to get $(V_p \cap Y) \subset (G \cap Y) = E$. Thus around each $p \in E$, there is a neighborhood of points in Y that is contained in E, so indeed E is open relative to Y.

To prove the other direction, we want to show that if E is open relative to Y, then we can build a set G that is open relative to X—where G must consist of E plus some other points of X. The basic idea is that we know every $p \in E$ has a neighborhood of points in Y that is completely contained in E. To build a set that is open relative to X, we need to "extend" those neighborhoods to include all points $\{x \in X \mid d(p, x) < r\}$, rather than just $\{y \in Y \mid d(p, y) < r\}$ (see Figure 10.3). How do we guarantee that all points in those extended neighborhoods are points of G? Easy—we just include them in G. Since all neighborhoods are open by Theorem 9.23, G will be a union of open sets in X, which Theorem 10.3 tells us is also open in X.

Let's prove this direction formally. If E is open relative to Y, then for each $p \in E$, there is an $r_p > 0$ such that all points $q \in Y$ with $d(p, q) < r_p$ are contained in E. Let V_p be the set of all points $q \in X$ with $d(p, q) < r_p$ (so V_p is the neighborhood "extended" to X), and let $G = \bigcup_{p \in E} V_p$. Each V_p is a neighborhood in X, so each V_p is open relative to X, so G is a union of open sets, so G is also open relative to X.

To prove that $E = Y \cap G$, we will use the method from Chapter 3 of showing $E \subset Y \cap G$ and $E \supset Y \cap G$. For every element p of E, $p \in E \implies p \in Y$, and $p \in V_p \implies p \in G$, so $p \in Y \cap G$. Thus $E \subset Y \cap G$. Also for every $p \in E$, by our construction of V_p, $V_p \cap Y$ is precisely the neighborhood of p in Y that is completely contained in E. Thus $E \supset \bigcup_{p \in E}(Y \cap V_p) = Y \cap \left(\bigcup_{p \in E} V_p\right)$, so $Y \cap G \subset E$. □

I should point out that as with openness, closedness is a relative property. A set E that is closed relative to Y may not be closed relative to X (if E has limit points that are in X but not Y).

Definition 10.15. *(Relative Closedness)*
A set E is **closed relative to** Y if, for each $p \in Y$ where

$$N_r(p) \cap E \neq \{\emptyset\} \text{ and } \neq \{p\},$$

then $p \in E$.

Example 10.16. *(Relative Closedness)*
If $X = \mathbb{R}$ and $Y = [0, 2)$, then $E = [1, 2)$ is closed in Y but not in X. (Because 2 is certainly a limit point of E, but since $2 \notin Y$, we don't need to have $2 \in E$ for E to be closed relative to Y.)

With a good understanding of open and closed sets under our belts, we can get into the real meat of topology: compact sets. (You thought we were done with new definitions for a while? Ha!)

CHAPTER 11

Compact Sets

Oh, compact sets. They can be very confusing for the first-timer, but they are an integral part of real analysis. (Get it? *Integral?!*) They come up many times later on, in relation to the properties of continuous functions.

We'll start by devoting this chapter to "compact set appreciation," in which we'll learn the definitions and get a sense of the cool properties that compact sets have. In the next chapter, we'll learn which types of sets in $\mathbb{R}$ are compact (and why all of this matters).

Definition 11.1. *(Open Cover)*
*For any subset E of a metric space X, an **open cover** of E in X is a collection of sets $\{G_\alpha\}$ that are open relative to X and contain E in their union.*

In symbols, $\{G_\alpha\}$ is an open cover of E if:

$$\forall \alpha, G_\alpha \text{ open relative to } X, \text{ and } E \subset \bigcup_\alpha G_\alpha.$$

*A **finite subcover** of an open cover $\{G_\alpha\}$ of E is a subcollection of $\{G_\alpha\}$ with a finite number of sets, which still contain E in their union.*

In symbols, $\{G_{\alpha_1}, G_{\alpha_2}, \ldots, G_{\alpha_n}\}$ is a finite subcover of $\{G_\alpha\}$ if:

$$n \in \mathbb{N}, \text{ and } E \subset \bigcup_{i=1}^n G_{\alpha_i}.$$

In the definition of a finite subcover, there are finitely many indexes $\alpha_1, \alpha_2, \ldots, \alpha_n$. Each G_{α_i} is an element of the collection $\{G_\alpha\}$. (To be completely formal, we should be writing an open cover as $\{G_\alpha \mid \alpha \in \mathcal{A}\}$, so a finite subcover has sets with indexes $\alpha_1, \alpha_2, \ldots, \alpha_n \in \mathcal{A}$.)

Example 11.2. *(Open Covers)*
The set

$$\{(z-2, z+2) \mid z \in \mathbb{Z}\}$$

is an open cover of the interval $[-3, 3]$, since each open interval $(z - 2, z + 2)$ (of length 4) is an open set. The set $\{(-4, 0), (-1, 3), (2, 6)\}$ is a finite subcover, along with $\{(-5, -1), (-3, 1), (-2, 2), (1, 5)\}$, and many other possible examples. As you can see, a single open cover can have multiple possible finite subcovers.

For a subset E of any metric space, choose any radius r_p for each $p \in E$ (the radii can be different). Then the collection $\{N_{r_p}(p) \mid p \in E\}$ is an open cover of E. Why? Obviously every point of E is included, and the union of neighborhoods is open by Theorem 10.3. (The proof of Theorem 10.14 uses a similar collection of sets.)

Let's say E contains a finite number of points. Then the open cover $\{N_{r_p}(p) \mid p \in E\}$ has a finite number of sets, so it is actually also a finite subcover (of itself). (Note that although each neighborhood could contain an infinite number of points, a finite subcover needs to have a finite number of *sets*, not points.)

Let's say E contains an infinite number of points, but we are told that $\{N_{r_p}(p) \mid p \in E\}$ contains a finite subcover. Then we know there must be a finite number of points $p_1, p_2, \ldots, p_n$ in E such that

$$E \subset \bigcup_{i=1}^{n} N_{r_{p_i}}(p_i).$$

Definition 11.3. *(Compact Set)*
*A subset K of a metric space X is **compact** if every open cover of K has a finite subcover. In symbols, K is compact if:*

$$\forall \text{ open cover } \{G_\alpha\} \text{ of } K, \exists \{\alpha_1, \alpha_2, \ldots, \alpha_n\} \text{ such that } K \subset \bigcup_{i=1}^{n} G_{\alpha_i}.$$

Compact sets are much more complex than open or closed sets. To show that a set is compact, we must show that *every* possible open cover has a finite subcover, which takes some major proving skills. Showing that a set is not compact is slightly easier, since we just need one example of an infinite open cover that has no finite subcover (but doing so can still be difficult).

Example 11.4. (Compact Sets)
The empty set is compact, since we can take one set G_{α_1} of any open cover $\{G_\alpha\}$, and $\emptyset \subset G_{\alpha_1}$.

Take a set with one point, $K = \{p\}$. Then K is clearly compact, since every open cover $\{G_\alpha\}$ must have at least one set that contains the point p. Take one of those sets that contains p, and we have found a finite subcover.

In fact, any set K with a finite number of points $p_1, p_2, \ldots, p_n$ is compact. Why? For any open cover $\{G_\alpha\}$, take a set $G_{\alpha_1} \in \{G_\alpha\}$ that contains p_1. Continue likewise: for each $p_i \in K$, if p_i is not contained in the sets we have taken so far, take a set $G_{\alpha_i} \in \{G_\alpha\}$ that contains p_i. After at most n steps, we have a finite collection $\{G_{\alpha_1}, G_{\alpha_2}, \ldots, G_{\alpha_n}\}$ of open sets whose union contains K.

This next theorem serves as a good basic example of a noncompact set.

Theorem 11.5. *(No Open Interval Is Compact)*
For any $a, b \in \mathbb{R}$ with $a < b$, no open interval (a, b) (or $(a, b) \cap \mathbb{Q}$) is compact.

Figure 11.1. Given any $x \in (a, b)$, we can find an N such that $x \in (a + \frac{1}{N}, b - \frac{1}{N})$. But given $(a + \frac{1}{N}, b - \frac{1}{N})$, we can find an element $y \in (a, b)$ outside it.

Proof. To show that (a, b) is not compact, we just need one example of an open cover that has no finite open subcover. We will use the cover

$$\{G_n\} = \left\{ \left(a + \frac{1}{n}, b - \frac{1}{n}\right) \;\middle|\; n \in \mathbb{N} \right\}.$$

(If this gives us an invalid interval such as $(1, 0)$, we just ignore that element.)

The key idea is drawn in Figure 11.1. We see that $\{G_n\}$ is a cover since given any element $x \in (a, b)$, we can use the Archimedean property to find an N large enough so that $x \in (a + \frac{1}{N}, b - \frac{1}{N})$. But $\{G_n\}$ has no finite subcover, since given an open interval $(a + \frac{1}{N}, b - \frac{1}{N})$, we can use the density property to find an element y outside that interval but $y \in (a, b)$.

Formally, why does $\bigcup_{n=1}^{\infty} G_n$ cover (a, b)? Well, for any element x with $a < x < b$, the Archimedean property provides an $N \in \mathbb{N}$ such that

$$N > \max\left\{\frac{1}{x - a}, \frac{1}{b - x}\right\}.$$

Then

$$Nx - Na > 1 \text{ and } Nb - Nx > 1 \implies Na + 1 < Nx < Nb - 1$$

$$\implies x \in \left(a + \tfrac{1}{N}, b - \tfrac{1}{N}\right).$$

Thus every element of (a, b) is in G_N for some $N \in \mathbb{N}$, so $(a, b) \subset \bigcup_{n=1}^{\infty} G_n$.

Formally, why does $\{G_n\}$ have no finite subcover? Well, for any $m > n$, we have

$$a + \frac{1}{m} < a + \frac{1}{n} < b - \frac{1}{n} < b - \frac{1}{m},$$

so that $G_m \supset G_n$. Thus for any finite $N \in \mathbb{N}$,

$$\bigcup_{n=1}^{N} G_n = G_N = \left(a + \tfrac{1}{N}, b - \tfrac{1}{N}\right).$$

But $(a, b) \not\subset G_N$—since, for example, $a + \frac{1}{2N} \in (a, b)$ but $a + \frac{1}{2N} \notin (a + \frac{1}{N}, b - \frac{1}{N})$. □

You might be asking, "if sets can be open and closed relative to some metric spaces but not others, what about compact sets?" In fact, in the definition of compactness, we

Compact Sets

completely sidestepped the issue. We didn't even specify whether we are dealing with open covers of K that are in X or in some other metric space.

As the next theorem shows, it really doesn't matter! A set K that is compact in one metric space is compact in *every* metric space (as long as that metric space contains K, of course).

Since a compact subset K of a metric space X is also a metric space and is compact anywhere, we will often refer to K as a *compact metric space*. Of course, using the phrase "closed metric space" or "open metric space" isn't really useful. Any metric space X is a closed subset of itself (since it contains all limit points in X) and an open subset of itself (since all neighborhoods contain only points of X).

Theorem 11.6. *(Relative Compactness)*
For any metric spaces $Y \subset X$, a subset K of Y is compact relative to X if and only if it is compact relative to Y.

Here, "K is compact relative to Y" means that for any open cover $\{V_\alpha\}$ with $V_\alpha \subset Y$ and V_α open relative to Y for each α, there is a finite subcover.

Proof. Let's start with the direction that assumes K is compact relative to X. The basic idea is that for any open cover of K in Y, we can expand it into an open cover of K in X and take a finite subcover in X, then intersect that back with Y to get a finite subcover in Y.

We want to show that for any arbitrary open cover $\{V_\alpha\}$ in Y there exists a finite subcover $\{V_{\alpha_1}, V_{\alpha_2}, \ldots, V_{\alpha_n}\}$ in Y. How do we relate sets that are open relative to one metric space, to those that are open relative to another? Use Theorem 10.14! So for every possible V_α, there exists a $G_\alpha \subset X$ that is open relative to X such that $V_\alpha = Y \cap G_\alpha$. Thus

$$K \subset \bigcup_\alpha V_\alpha = \bigcup_\alpha (Y \cap G_\alpha) = Y \cap \left(\bigcup_\alpha G_\alpha \right).$$

We already know K is a subset of Y, so the above line shows that $\{G_\alpha\}$ is an open cover of K. Because K is compact relative to X, there is a finite subcover $\{G_{\alpha_1}, G_{\alpha_2}, \ldots, G_{\alpha_n}\} \subset \{G_\alpha\}$. Each G_{α_i} is an element of $\{G_\alpha\}$, so for every i between 1 and n, $V_{\alpha_i} = Y \cap G_{\alpha_i}$, where $V_{\alpha_i} \in \{V_\alpha\}$. Thus

$$K \subset Y \text{ and } K \subset \bigcup_{i=1}^n G_{\alpha_i}$$

$$\implies K \subset Y \cap \left(\bigcup_{i=1}^n G_{\alpha_i} \right) = \bigcup_{i=1}^n (Y \cap G_{\alpha_i}) = \bigcup_{i=1}^n V_{\alpha_i}.$$

This shows that $\{V_\alpha\}$ has a finite subcover of K, so K is compact relative to Y.

The other direction uses the same argument in reverse. Try filling in the blanks in Box 11.1.

BOX 11.1

PROVING THE SECOND DIRECTION OF THEOREM 11.6

Assume K is compact relative to Y, and let $\{G_\alpha\}$ be an open cover of K in X. Let $V_\alpha = Y \cap \underline{\hspace{1cm}}$, so by Theorem 10.14, V_α is open $\underline{\hspace{2cm}}$. Because $K \subset \bigcup_\alpha G_\alpha$ and $K \subset Y$, we have $K \subset \underline{\hspace{2cm}} = \bigcup_\alpha V_\alpha$, so there is a finite subcover $\{V_{\alpha_i}\}$ of K in Y. Then

$$K \subset \bigcup_{i=1}^n V_{\alpha_i} = \bigcup_{i=1}^n \underline{\hspace{1cm}} = \underline{\hspace{1cm}} \cap \underline{\hspace{1cm}},$$

so $\{G_{\alpha_i}\}$ is a finite subcover of K in $\underline{\hspace{1cm}}$.

□

The next few theorems help us better characterize which sets are compact by relating compactness to closedness.

Theorem 11.7. *(Compact Subsets Are Closed)*
For any compact subset K of a metric space X, K is closed in X.

Because a compact set K in X is compact in any metric space, this theorem implies that any compact set is closed in *every* possible metric space (that it can be a subset of).

Notice that if we take the contrapositive of this theorem, we have "if K is not closed relative to *some* metric space X, then K is not compact in any metric space." We could have used this to prove Theorem 11.5 more easily: (a, b) is not closed in $\mathbb{R}$ (since a and b are limit points), so (a, b) is not compact.

Proof. It will be simpler to prove that K^C is an open subset of X (and then apply Theorem 10.1). We want to show that there is a neighborhood around any point p of K^C that is contained in K^C. We can do this by covering K in neighborhoods around its points (making sure the neighborhoods don't contain p) and then taking a finite subcover of K. Now a neighborhood around p of radius less than the distance to the nearest open set of the cover will be contained in K^C.

To make this explicit, take an arbitrary $p \in X$ with $p \notin K$ (so $p \in K^C$). We want to show that there is some $r > 0$ such that $N_r(p) \subset K^C$.

We saw in Example 11.2 that the set of neighborhoods (of any radius) around each point of K is an open cover of K. For any point $q \in K$, let

$$W_q = N_{\frac{1}{3}d(p,q)}(q),$$

so that $\{W_q \mid q \in K\}$ is an open cover of K (see Figure 11.2). Because K is compact, this open cover must have a finite subcover, so there must be finitely many points $q_1, q_2, \ldots, q_n$ such that $K \subset \bigcup_{i=1}^n W_{q_i}$.

Because $\{q_1, q_2, \ldots q_n\}$ is a finite set, we can choose the element that is closest to p. Let

$$d = \min_{1 \le i \le n} \{d(p, q_1), d(p, q_2), \ldots d(p, q_n)\},$$

and let $V = N_{\frac{1}{3}d}(p)$.

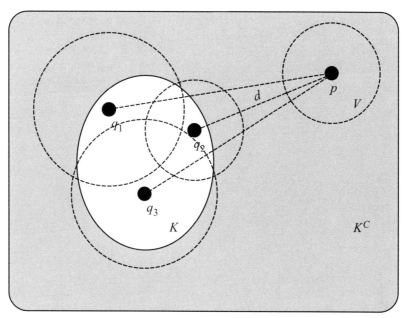

Figure 11.2. The set K in this figure is covered by the neighborhoods of three points, each having radius $\frac{1}{3}d(p, q_i)$. The number d is the smallest distance between p and any q_i. The neighborhood V around p of radius $\frac{1}{3}d$ does not intersect with a neighborhood of any q_i, so it is completely contained in K^C.

Then for any i between 1 and n,

$$\frac{1}{3}d < \frac{1}{3}d + \frac{1}{3}d(p, q_i)$$
$$= \frac{1}{3}d(p, q_i) + \frac{1}{3}\min_{1 \leq i \leq n}\{d(p, q_1), d(p, q_2), \ldots d(p, q_n)\}$$
$$\leq \frac{1}{3}d(p, q_i) + \frac{1}{3}d(p, q_i)$$
$$< d(p, q_i),$$

so V and W_{q_i} do not intersect.

Thus $V \cap \left(\bigcup_{i=1}^{n} W_{q_i}\right) = \emptyset$, so $V \cap K = \emptyset$. Since V has no point in common with K, it must be entirely contained in K^C, so p is indeed an interior point of K^C. □

 Notice that this proof hinges on the fact that a finite subcover of K exists—if it did not, we would not be able to take the q_i with minimum distance from p (since minimums are not necessarily defined for infinite sets).

Also note that our open cover and finite subcover depended on p. Knowing a set is compact lets us be manipulative in this way, since *every* open cover has a finite subcover, so we can pick and choose according to our needs.

Theorem 11.8. *(Closed Subsets of Compact Sets Are Compact)*
For any metric spaces $K \subset X$, if K is compact and if a subset F of K is closed relative to X, then F is also compact.

Proof. The proof is actually pretty simple—the key point being that the complement of F in K is open, so we can just adjoin it to an open cover of F to get an open cover of K, then take a finite subcover and remove F^C.

Let $\{V_\alpha\}$ be an open (relative to K) cover of F. Since $K = F \cup F^C$ (here F^C denotes the complement of F in K, not in X) and $F \subset \bigcup_\alpha V_\alpha$, we have

$$K \subset \left(\bigcup_\alpha V_\alpha\right) \cup F^C.$$

Each V_α is open (relative to K), and so is F^C, so $\{V_\alpha\} \cup F^C$ is an open cover of K. Because K is compact, there is a finite subcover $\{V_{\alpha_1}, V_{\alpha_2}, \ldots, V_{\alpha_n}, F^C\}$ (where F^C may or may not need to be an element of this finite subcover).

Then we have

$$F \subset K \subset \left(\bigcup_{i=1}^n V_{\alpha_i}\right) \cup F^C,$$

and since $F \cap F^C = \emptyset$, this implies that $\{V_{\alpha_1}, V_{\alpha_2}, \ldots, V_{\alpha_n}\}$ is a finite subcover of F. $\square$

Corollary 11.9. *(Intersection of Closed Sets and Compact Sets)*
For any metric spaces $K \subset X$, if K is compact and if a subset F of X is closed relative to X, then $F \cap K$ is also compact.

This corollary tells us that if a compact set K is embedded in any metric space, the intersection of K and any closed set in that metric space is compact.

Proof. By Theorem 11.7, K is closed in X. Since the intersection of closed sets is closed, we have that $F \cap K$ is closed. Notice that $F \cap K \subset K$, so by Theorem 11.8, $F \cap K$ is compact. $\square$

It's important to understand that compactness is a universal property, whereas closedness depends on the embedding metric space. Maybe some of these distinctions have become muddled throughout the last few theorems, so here's a little test: what is wrong with the following paragraph? (We will later learn that any closed interval $[a, b]$ is compact; take this fact as a given for now.)

Let X be the metric space given by the open interval $(-3, 3)$, and let $F = X$. Then F is closed in X (since $p \in X \implies p \in F$, so F contains all of its limit points). Let $K = [-5, 5]$, so K is compact. Then by Corollary 11.9, $F \cap K = (-3, 3)$ is compact, which contradicts Theorem 11.5!

Did you find the error? This is a blatant abuse of Corollary 11.9. In the example, we had $F \subset X \subset \mathbb{R}$ and $K \subset \mathbb{R}$. F was closed in X but *not* in $\mathbb{R}$, so all Corollary 11.9 tells us is that for any compact set K in X, $F \cap K$ is compact. But $K = [-5, 5]$ is not a subset of X, so we cannot apply the corollary. (We could, however, take something like $K = [-2, 2]$ to conclude that $F \cap K = [-2, 2]$ is compact.)

Theorem 11.10. *(Compact Sets Are Bounded)*
For any compact subset K of a metric space X, K is bounded in X.

Proof. To show *K* compact $\implies$ *K* bounded, we'll do a proof by contrapositive (we'll show *K* not bounded $\implies$ *K* not compact). The definition of boundedness is

$$\exists q \in X, \exists M \in \mathbb{R} \text{ such that } d(p, q) \leq M, \forall p \in K,$$

so the opposite is

$$\forall q \in X, \forall M \in \mathbb{R}, \exists p \in K \text{ such that } d(p, q) > M.$$

If *K* were compact, then the open cover $\{N_1(p) \mid p \in K\}$ (neighborhoods of radius 1 around each point of *K*) would have a finite subcover $\{N_1(p_1), N_1(p_2), \ldots, N_1(p_n)\}$. Let's take the furthest distance from p_1, calling it

$$r = \max \{d(p_1, p_i) \mid 1 \leq i \leq n\}.$$

Then $N_{r+1}(p_1)$ contains that finite subcover $\{N_1(p_1), N_1(p_2), \ldots, N_1(p_n)\}$, since for any *i* between 1 and *N*, $d(p_1, p_i) + 1 \leq r + 1$, so $N_1(p_i) \subset N_{r+1}(p_1)$. Now because *K* is not bounded, for $q = p_1$ and $M = r + 1$, there is some $p \in K$ such that $d(p, p_1) > r + 1$, so that $p \notin N_{r+1}(p_1)$. Thus at least one element of *K* is not in a superset of the finite subcover, so it is not in the finite subcover, so *K* cannot be compact. $\square$

Theorem 11.11. *(Intersection of Compact Sets)*
Let $\{K_\alpha\}$ be a collection of nonempty compact sets in a metric space X. If the intersection of every finite subcollection of $\{K_\alpha\}$ is nonempty, then so is $\bigcap_\alpha K_\alpha$.

Isn't this a trivial property? If the intersection of every possible combination of sets in $\{K_\alpha\}$ is nonempty, how could the intersection of all of the sets be empty?

This can happen because of the trickiness of infinite intersections. For example, let $A_n = (-\frac{1}{n}, \frac{1}{n})$, and take the the infinite collection of sets $\{A_n \mid n \in \mathbb{N}\}$. We know that $\bigcap_{n=1}^\infty A_n = \{0\}$ (because for any nonzero point *p*, there exists an open interval $(-\frac{1}{k}, \frac{1}{k})$ such that either $p < -\frac{1}{k}$ or $p > \frac{1}{k}$).

Now define $B_n = A_n \setminus \{0\} = (-\frac{1}{n}, 0) \cup (0, \frac{1}{n})$. Then $\bigcap_{n=1}^\infty B_n = \emptyset$, so there is no element that belongs to every B_n. But every *finite* collection of $\{B_n\}$ *does* have at least one element in common. Why? For any $m > n$, $B_m \subset B_n$, so

$$\bigcap_{n=1}^k B_k = B_k = \left(-\tfrac{1}{k}, 0\right) \cup \left(0, \tfrac{1}{k}\right) \neq \emptyset.$$

Theorem 11.11 basically says that what happened with this collection $\{B_n\}$ cannot happen with compact sets. (In this case, each B_n is not closed in $\mathbb{R}$ and therefore not compact.)

Proof. We'll prove this by contradiction. If $\bigcap_\alpha K_\alpha$ is empty, then one set K_1 can be covered by a finite number of sets K_α^C, but a finite intersection of sets K_α is empty.
To make this more clear, we will use the following property of sets:

$$A \cap \left(\bigcap_\alpha B_\alpha\right) = \emptyset \iff A \subset \bigcup_\alpha B_\alpha^C.$$

To prove one direction of the implication, take $x \in A$. Then $x \notin B_\alpha$ for at least one α—otherwise x would be an element of every B_α, so $A \cap \left(\bigcap_\alpha B_\alpha\right)$ would contain x, not be empty—so $x \in B_\alpha^C$ for at least one α. Thus every element of A is in some B_α^C, so $A \subset \bigcup_\alpha B_\alpha^C$.

To prove the other direction, we assume that if $x \in A$, then $x \in B_\alpha^C$ for some α. So every element of A is missing from at least one B_α, so there cannot be anything in both A and every B_α. Thus $A \cap \left(\bigcap_\alpha B_\alpha\right) = \emptyset$.

Now pick a compact set K_1 from the collection. Assume that $\bigcap_\alpha K_\alpha$ is empty, so $K_1 \cap \left(\bigcap_{\alpha \neq 1} K_\alpha\right) = \emptyset$. By the property above, this means that $K_1 \subset \bigcap_{\alpha \neq 1} K_\alpha^C$. Also, because Theorem 11.7 says that each K_α is closed relative to X, each K_α^C must be open relative to X. Thus $\{K_\alpha^C \mid \alpha \neq 1\}$ is an open cover of K_1.

Well K_1 is compact, so there is a finite subcover $\{K_{\alpha_1}^C, K_{\alpha_2}^C, \ldots, K_{\alpha_n}^C\}$ of K_1. Remember $K_1 \subset \bigcup_{i=1}^n K_{\alpha_i}^C$ implies

$$K_1 \cap \left(\bigcap_{i=1}^n K_{\alpha_i}\right) = \emptyset.$$

Thus a finite intersection of K_α's is empty, which contradicts an assumption of the theorem. So $\bigcap_\alpha K_\alpha$ cannot be empty. $\square$

We will end our (seemingly endless) rambling of the awesome powers of compact sets with one last corollary and one last theorem. Everything up to now has built on itself, so we can finally come up with some results that you will (hopefully) find interesting and useful.

Corollary 11.12. *(Nested Compact Sets)*
Let $\{K_n\}$ be a collection of nonempty compact sets such that for any $n \in \mathbb{N}$, $K_n \supset K_{n+1}$. Then the intersection $\bigcap_{n=1}^\infty K_n \neq \emptyset$.

These sets are called "nested" because each set in the sequence contains all subsequent sets in the sequence. Think of the collection as a Russian nesting doll: you open up K_1 (Olga) to find a smaller doll K_2 (Galina) inside, which contains K_3 (Anastasia), and so forth. The corollary guarantees that this collection of dolls—which you can keep cracking open to find a smaller doll, forever—has *something* inside of them. (The next time you visit a souvenir shop in Russia, show off your math skills by talking about compactness. *"You know compact sets? You get half price!"*)

Proof. For any $m > n$, we have $K_m \subset K_n$, so we know that

$$\bigcap_{n=1}^k K_n = K_k \neq \emptyset.$$

Thus the intersection of any finite subcollection of $\{K_n\}$ is nonempty, so by Theorem 11.11, $\bigcap_{n \in \mathbb{N}} K_n \neq \emptyset$. $\square$

Theorem 11.13. *(Limits in Compact Sets)*
For any infinite subset E of a compact set K, E has at least one limit point in K.

The key prerequisite is that E has an infinite number of points. (Of course, if E were finite, the theorem couldn't possibly be true. E would have no limit points at all, since Theorem 9.11 tells us that any neighborhood around a limit point of E contains an infinite number of points of E.)

Proof. Assume the result is false—namely, E has no limit points in K. Then for every $q \in K$, there exists some $r_q > 0$ such that $N_{r_q}(q) \cap E = \{q\}$ or $= \emptyset$. The collection $\{N_{r_q}(q) \mid q \in K\}$ is an open cover of K, and each set in the collection contains at most one point of E.

Because K is compact, that open cover has a finite subcover

$$\{N_{r_{q_1}}(q_1), N_{r_{q_2}}(q_2), \ldots, N_{r_{q_n}}(q_n)\}.$$

But that finite subcover only contains a finite number of points of E—specifically, it contains $\{q_1, q_2, \ldots, q_n\} \cap E$—while E has an infinite number of points. Thus E is not contained in that subcover, so neither is K, which is a contradiction. □

You'll notice that except for a few examples early on, we haven't considered which sets are actually compact. We now know what they can do, we just don't know how to find them.

As promised, the next chapter will prove that every interval $[a, b]$ is compact. We'll also go over the Heine-Borel theorem, a central result in real analysis that tells us how to find compact sets in $\mathbb{R}^k$.

CHAPTER 12

The Heine-Borel Theorem

If you like surprises, then skip this paragraph; otherwise, spoiler alert! I'm going to tell you what the Heine-Borel theorem is: any subset of $\mathbb{R}^k$ is compact if and only if it is both closed and bounded. The last chapter showed us that every compact set (in any metric space, not just $\mathbb{R}^k$) is closed and bounded, so the real meat of the Heine-Borel theorem is the reverse implication. It asserts that every closed and bounded set in $\mathbb{R}^k$ is compact, which is not true in every metric space. For example, the set $(-\pi, \pi) \cap \mathbb{Q}$ is closed in $\mathbb{Q}$ (since $-\pi$ and π are not rational) and is bounded but it is not compact (by Theorem 11.5).

Watch out! This chapter has a few lengthy proofs, which may look difficult to the untrained eye. (But who are we kidding, your eye is definitely trained by now.) As you read long passages, it can be very helpful to draw pictures in the margins to clarify what is going on at each step.

The last chapter was concerned with compactness in arbitrary metric spaces X, but this chapter cares about $\mathbb{R}^k$. To discover which sets in $\mathbb{R}^k$ are compact (remember from Chapter 6 that $\mathbb{R}^k$ is the set of all k-dimensional vectors of real numbers), we'll start by looking at a common property of both intervals and compact sets.

Theorem 12.1. *(Nested Closed Interval Property)*
Let $\{I_n\}$ be a collection of closed intervals in $\mathbb{R}$ such that for any $n \in \mathbb{N}$, $I_n \supset I_{n+1}$. Then $\bigcap_{n=1}^{\infty} I_n \neq \emptyset$.

This theorem should look very familiar, since we proved the analogous property for compact sets in Corollary 11.12. The nested closed interval property is an inherent feature of $\mathbb{R}$, which follows as an immediate consequence of the least upper bound property.

Proof. Each interval is of the form $I = [a_n, b_n]$. We want to find a number x such that $x \in [a_n, b_n]$ for all $n \in \mathbb{N}$. (If this is possible, we'll have $\bigcap_{n=1}^{\infty} I_n \supset \{x\}$.) The supremum x of the set of all upper bounds a_n seems like a good choice, since it should be $\geq a_n$ and $\leq b_n$ for any $n \in \mathbb{N}$, as we can see in Figure 12.1.

Let $E = \{a_n \mid n \in \mathbb{N}\}$. Any b_n is an upper bound of E—since if $a_n > b_m$ for some $n, m \in \mathbb{N}$, then $I_n \cap I_m = \emptyset$, so neither interval can contain the other, which is a contradiction. Thus E is a nonempty subset of $\mathbb{R}$ that is bounded above, so by the least upper bound property of $\mathbb{R}$, $\sup E$ exists in $\mathbb{R}$. Let's call it x.

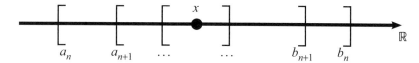

Figure 12.1. The point $x = \sup\{a_n \mid n \in \mathbb{N}\}$ is $\geq a_n$ and $\leq b_n$ for any $n \in \mathbb{N}$.

Now for any $n \in \mathbb{N}$, $x \geq a_n$ since x is an upper bound of E. Moreover, $x \leq b_n$ since b_n is an upper bound of E, and x is the *least* upper bound. Thus $x \in [a_n, b_n]$.

Note that the same proof would have also worked if we had let $x = \inf\{b_n \mid n \in \mathbb{N}\}$. □

In fact, this property holds not just for intervals in $\mathbb{R}$ but also for the equivalent idea of intervals in $\mathbb{R}^k$, which we call k-cells.

Definition 12.2. *(k-cells)*
If $a_j < b_j$ for every j between 1 and k, then the set of vectors in $\mathbb{R}^k$ given by

$$\{\mathbf{x} = (x_1, x_2, \ldots, x_k) \mid x_j \in [a_j, b_j], \forall 1 \leq j \leq k\}$$

*is called a **k-cell**.*

We can visualize k-cells in Figure 12.2. In one dimension, we set $k = 1$, and see that a 1-cell is just a closed interval. In two dimensions, a 2-cell is a rectangle, since it is the set of all points that fall between two sets of bounds. In three dimensions, a 3-cell is a box. In four dimensions—well, according to Einstein, the fourth dimension might be time, so I guess a 4-cell is a box that forever remains in some fixed period of time (to see an example, visit your grandparents' attic).

Theorem 12.3. *(Nested k-cell Property)*
Let $\{I_n\}$ be a collection of k-cells in $\mathbb{R}^k$ such that for any $n \in \mathbb{N}$, $I_n \supset I_{n+1}$. Then $\bigcap_{n=1}^{\infty} I_n \neq \emptyset$.

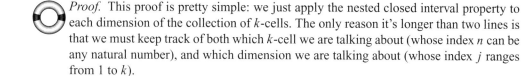

Proof. This proof is pretty simple: we just apply the nested closed interval property to each dimension of the collection of k-cells. The only reason it's longer than two lines is that we must keep track of both which k-cell we are talking about (whose index n can be any natural number), and which dimension we are talking about (whose index j ranges from 1 to k).

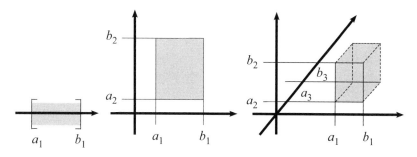

Figure 12.2. A 1-cell, 2-cell, and 3-cell (make sure to put on your 3D glasses for the optimal viewing experience).

Each k-cell I_n is the set of points $\mathbf{x} = (x_1, x_2, \ldots, x_k)$ such that $a_{n_j} \leq x_j \leq b_{n_j}$ for every j between 1 and k. Thus I_n is the set of vectors whose jth coordinate lies in the closed interval $I_{n_j} = [a_{n_j}, b_{n_j}]$. Since each k-cell is contained in the previous one, we have $I_{n_j} \supset I_{(n+1)_j}$ for every $n \in \mathbb{N}$ and every j between 1 and k.

We want to show that there is some vector $\mathbf{x} = (x_1, x_2, \ldots, x_k)$ that belongs to every I_n, meaning the vector's coordinates satisfy $x_1 \in I_{n_1}$ for every $n \in \mathbb{N}$, $x_2 \in I_{n_2}$ for every $n \in \mathbb{N}$, and ... and $x_k \in I_{n_k}$ for every $n \in \mathbb{N}$.

Well, by Theorem 12.1, the intersection $\bigcap_{n=1}^{\infty} I_{n_1}$ contains at least one point, which we'll call x_1^*. Similarly, there is some $x_2^* \in \bigcap_{n=1}^{\infty} I_{n_2}$, and ... and some $x_k^* \in \bigcap_{n=1}^{\infty} I_{n_k}$. Thus the vector $\mathbf{x}^* = (x_1^*, x_2^*, \ldots, x_k^*)$ is in I_n for every $n \in \mathbb{N}$. □

We will be able to prove not only that every closed interval is compact but in fact that every k-cell is compact.

The previous theorem will be useful for our proof, but note that it is not a proof in and of itself. (Just because k-cells have one of the same properties as compact sets, does not necessarily mean that all k-cells are compact.)

Theorem 12.4. *(k-cells Are Compact)*
For any $k \in \mathbb{N}$, every k-cell is a compact set.

Proof. Although this might look like a complicated argument, the underlying logic isn't too bad. We'll take it in two steps. First let's figure out what we want to say, then write it up formally.

Step 1. For a proof by contradiction, assume that a k-cell I is not compact, so some open cover $\{G_\alpha\}$ has no finite subcover. We then divide I into a finite number of smaller k-cells. At least one of these sub-k-cells cannot be covered by any finite subcollection of $\{G_\alpha\}$ (otherwise $\{G_\alpha\}$ would have a finite subcover for all of I), so take that sub-k-cell and divide it into smaller k-cells. Similarly, at least one of those sub-sub-k-cells cannot be finitely covered.... We keep splitting up, as in Figure 12.3, to obtain a collection of nested k-cells, none of which can be finitely covered.

By Theorem 12.3, their intersection must contain at least one element $\mathbf{x}^*$. Since $\mathbf{x}^* \in I$, there must be some set G_1 in $\{G_\alpha\}$ that contains it. But because G_1 is open, some neighborhood around $\mathbf{x}^*$ is also contained in G_1. Since each of those nested sub-k-cells contains a smaller sub-k-cell, we can find one arbitrarily small enough to fit inside that neighborhood of $\mathbf{x}^*$. Thus one of the nested sub-k-cells is covered by G_1—a finite subcollection of $\{G_\alpha\}$—which gives us a contradiction.

For that last step of the argument to work, we must keep track of how big these sub-k-cells are. Here's how we'll do the division: the k-cell I is given by an interval in each dimension, so let's divide each of those intervals in half, giving us 2^k sub-k-cells (as you can see in Figure 12.3, in two dimensions this means we divide I into $2^2 = 4$ parts). At each step of the division process, we divide the sub-k-cell that cannot be finitely covered into 2^k more sub-k-cells, so after n steps, we have a sub-k-cell that is $\left(\frac{1}{2^k}\right)\left(\frac{1}{2^k}\right)\cdots\left(\frac{1}{2^k}\right) = 2^{-nk}$ times the volume of the original k-cell I.

How big is I? We are concerned about the distance between two points in a k-cell (remember, we want it to be less than the radius of the neighborhood around $\mathbf{x}^*$ in G_1). This distance is limited by the largest possible distance within I, which is given by its diagonal. Let's put the ends of the k closed intervals $[a_j, b_j]$, which make up I into coordinates of a vector, to get $\mathbf{a} = (a_1, a_2, \ldots, a_n)$ and $\mathbf{b} = (b_1, b_2, \ldots, b_n)$.

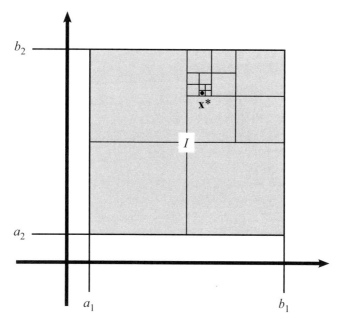

Figure 12.3. An example for a 2-cell I, which we divide into four sub-2-cells at each step, then select the sub-2-cell that is not finitely covered, which we divide again. Because this is a sequence of nested 2-cells, their intersection contains at least one point $\mathbf{x}^*$.

The diagonal is then given by

$$|\mathbf{b} - \mathbf{a}| = + \left(\sum_{i=1}^{k} (b_j - a_j)^2 \right)^{\frac{1}{2}}$$

(using Definition 6.10), which we label δ. So the distance between any two points in the nth sub-k-cell is $\leq 2^{-n}\delta$.

We want to find a sub-k-cell small enough to be contained in the neighborhood of $\mathbf{x}^*$ in G_1. So for $N_r(\mathbf{x}^*) \subset G_1$, if we can show that $2^{-n}\delta < r$, then the distance between $\mathbf{x}^*$ and any point of the nth sub-k-cell is less than r—meaning the nth sub-k-cell is contained in $N_r(\mathbf{x}^*)$ and thus also in G_1. We want $\delta < 2^n r$, meaning

$$n > \frac{\log(\frac{\delta}{r})}{\log(2)}.$$

Because $\delta > 0$ and $r > 0$, we also have $\log(\frac{\delta}{r}) \in \mathbb{R}$. Then by the Archimedean property of $\mathbb{R}$, such an $n \in \mathbb{N}$ does exist.

Step 2. Here is the formal proof.

Fix a dimension k and let I be a k-cell, so

$$I = \{\mathbf{x} = (x_1, x_2, \ldots, x_k) \mid a_j \leq x_j \leq b_j, \forall 1 \leq j \leq k\},$$

and let

$$\delta = \left(\sum_{i=1}^{k}(b_j - a_j)^2\right)^{\frac{1}{2}}.$$

Then for any $\mathbf{x}, \mathbf{y} \in I$, $|\mathbf{x} - \mathbf{y}| \leq \delta$.

Assume I is not compact, then choose an open cover $\{G_\alpha\}$ of I which has no finite subcover. Split I into 2^k sub-k-cells Q_i by splitting $[a_j, b_j]$ into

$$\left[a_j, \frac{a_j + b_j}{2}\right] \text{ and } \left[\frac{a_j + b_j}{2}, b_j\right],$$

so that $\bigcup_{i=1}^{2^k} Q_i = I$. There must be at least one set $I_1 \in \{Q_i \mid 1 \leq i \leq 2^k\}$ which is not covered by any finite subcollection of $\{G_\alpha\}$. Then split I_1 into 2^k sub-k-cells, and continue the process to create a collection of k-cells $\{I_n\}$ with $I \supset I_1 \supset I_2 \supset \ldots$, where no I_n is finitely covered.

By Theorem 12.3, there is some $\mathbf{x}^* \in I_n$ for every $n \in \mathbb{N}$. Also, $\mathbf{x}^* \in I$, so there must be some $G_1 \in \{G_\alpha\}$ such that $\mathbf{x}^* \in G_1$. Since G_1 is open, there exists some $r > 0$ such that $N_r(\mathbf{x}^*) \subset G_1$. By the Archimedean property of $\mathbb{R}$, there exists some $n \in \mathbb{N}$ such that $n > \frac{\log(\frac{\delta}{r})}{\log(2)}$, so that $2^{-n}\delta < r$. Note that for any $\mathbf{y} \in I_n$, $|\mathbf{x}^* - \mathbf{y}| \leq 2^{-n}\delta < r$, so $I_n \subset N_r(\mathbf{x}^*) \subset G_1$. Thus some finite subcollection of $\{G_\alpha\}$ (namely, the one-element subcollection consisting of G_1) covers I_n, which is a contradiction. Therefore, I must be compact. □

We are almost ready to prove the Heine-Borel theorem, which characterizes which sets in $\mathbb{R}^k$ are compact. First, though, we'll prove the following result, which will come in handy.

Theorem 12.5. *(Bounded Sets in $\mathbb{R}^k$)*
If a subset E of $\mathbb{R}^k$ is bounded, then it is contained in some k-cell.

Review Definition 9.3 to make sure you remember what being bounded in a metric space means.

Proof. The basic idea is that if every point of E is less than M away from some $\mathbf{q} \in \mathbb{R}^k$, we can build a k-cell based on that bound M and $\mathbf{q}$. (In geometric terms, every circle can be inscribed in a square, every sphere can be inscribed in a cube, etc.)

To make this precise, if E is bounded, then there is some $\mathbf{q} \in \mathbb{R}^k$ such that for every point $\mathbf{p}$ of E, $|\mathbf{p} - \mathbf{q}| \leq M$ for some $M \in \mathbb{R}$. If we write $\mathbf{p} = (p_1, p_2, \ldots, p_k)$ and $\mathbf{q} = (q_1, q_2, \ldots, q_k)$, this means

$$\left(\sum_{i=1}^{k}(p_j - q_j)^2\right)^{\frac{1}{2}} \leq M.$$

Then for each j between 1 and k, we have

$$0+\ldots+0+(p_j-q_j)^2+0+\ldots+0 \leq (p_1-q_1)^2+(p_2-q_2)^2+\ldots+(p_k-q_k)^2 \leq M^2,$$

so that $q_j - M \leq p_j \leq q_j + M$. Remember this is true for every $\mathbf{p} \in E$, so if we let

$$I = \left\{\mathbf{x} = (x_1, x_2, \ldots, x_k) \mid x_j \in [q_j - M, q_j + M], \forall 1 \leq j \leq k\right\},$$

then I is a k-cell, and $E \subset I$. □

Theorem 12.6. *(The Heine-Borel Theorem)*
A subset E of $\mathbb{R}^k$ is compact if and only if it is closed and bounded.

Here "closed and bounded" means closed in $\mathbb{R}^k$ and bounded in $\mathbb{R}^k$.

Proof. We've already done most of the work with all the theorems we've been proving in this chapter.

If E is compact, then by Theorem 11.7, E is closed in $\mathbb{R}^k$, and by Theorem 11.10, E is bounded in $\mathbb{R}^k$.

To prove the other direction of the theorem, assume E is closed and bounded. Because it is bounded, by Theorem 12.5, it must be contained in some k-cell I. By Theorem 12.4, I is compact. And since $E \subset I$ and E is closed, E is compact by Theorem 11.8. □

In addition to the Heine-Borel theorem, the following theorem also gives another way of finding compact sets in $\mathbb{R}^k$. Remember that Theorem 11.13 showed that every infinite subset of a compact set has a limit point in the compact set. We will now prove the converse in $\mathbb{R}^k$.

Theorem 12.7. *(Limits in Real Compact Sets)*
A subset E of $\mathbb{R}^k$ is compact if and only if every infinite subset of E has a limit point in E.

Proof. Assume E is compact. Then every infinite subset of E has a limit point in E by Theorem 11.13.

Conversely, assume every infinite subset of E has a limit point in E. We want to show E is compact, so by the Heine-Borel theorem, we just need to show that E is closed and bounded. Let's prove the contrapositive: if E is *not* bounded, then some infinite subset of E has no limit point in E; similarly, if E is *not* closed, then some infinite subset of E has no limit point in E.

If E is not bounded, we can construct an infinite subset S of E with no limit point in E. For $\mathbf{q} = \mathbf{0}$ and $M = 1, 2, 3, \ldots$, there are points $\mathbf{x}_n \in E$ such that $|\mathbf{x}_n - \mathbf{q}| > M$. So for every $n \in \mathbb{N}$, there is some $\mathbf{x}_n \in E$ with $|\mathbf{x}_n| > n$. Let $S = \{\mathbf{x}_n \mid n \in \mathbb{N}\}$, so $S \subset E$.

Also, S is infinite. If S had a finite number of points $\{\mathbf{x}_1, \mathbf{x}_2, \ldots, \mathbf{x}_N\}$, then we could take

$$n = \lceil \max\{|\mathbf{x}_1|, |\mathbf{x}_2|, \ldots, |\mathbf{x}_N|\} \rceil + 1$$

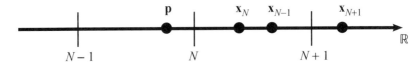

Figure 12.4. An example of points of S in $k = 1$ dimension, where N is the smallest integer $\geq |\mathbf{p}|$. Notice that for example, $\mathbf{x}_{N-1}$ could be greater than $\mathbf{x}_N$. But the important thing is that for any $n > N$, $|\mathbf{x}_n - \mathbf{p}| > 1$.

(using the ceiling notation as in Example 4.8 to mean "round up"). So there would be no $\mathbf{x}_n \in E$ with $|\mathbf{x}_n| > n$, which is a contradiction.

Furthermore, no point $\mathbf{p} \in \mathbb{R}^k$ is a limit point of S. Why? See Figure 12.4. Given any $\mathbf{p} \in \mathbb{R}^k$, let N be the smallest natural number $\geq |\mathbf{p}|$. Then for any natural number n greater than N, we have $n \geq N + 1$, so that

$$|\mathbf{x}_n - \mathbf{p}| \geq |\mathbf{x}_n| - |\mathbf{p}| \quad \text{(by Property 6 of Theorem 6.11)}$$
$$> n - |\mathbf{p}|$$
$$\geq N + 1 - |\mathbf{p}|$$
$$\geq 1 \quad \text{(since } N \geq |\mathbf{p}|\text{)}.$$

Now there are two cases. If $\mathbf{p} \notin S$, then $\mathbf{p} \neq \mathbf{x}_n$ for any $n \in \mathbb{N}$, so let

$$r = \frac{1}{2} \min \left\{ |\mathbf{x}_1 - \mathbf{p}|, |\mathbf{x}_2 - \mathbf{p}|, \ldots, |\mathbf{x}_N - \mathbf{p}|, 1 \right\}.$$

Then for any $n \in \mathbb{N}$, if $n \leq N$ we have $|\mathbf{x}_n - \mathbf{p}| > r$, and if $n > N$, we have $|\mathbf{x}_n - \mathbf{p}| > 1 > r$. So $N_r(\mathbf{p})$ does not contain $\mathbf{x}_n$ for any $n \in \mathbb{N}$, so $N_r(\mathbf{p}) \cap S = \emptyset$, so $\mathbf{p}$ is not a limit point of S.

If $\mathbf{p} \in S$, then there exists some $i \leq N$ such that $\mathbf{p} = \mathbf{x}_i$ (since remember, $N \geq |\mathbf{p}|$). Let

$$r = \frac{1}{2} \min \left\{ |\mathbf{x}_1 - \mathbf{p}|, |\mathbf{x}_2 - \mathbf{p}|, \ldots, |\mathbf{x}_{i-1} - \mathbf{p}|, |\mathbf{x}_{i+1} - \mathbf{p}|, \ldots |\mathbf{x}_N - \mathbf{p}|, 1 \right\}$$

(where we have removed $\mathbf{p} = \mathbf{x}_i$ from that minimum). By the same logic as in the first case, $N_r(\mathbf{p}) \cap S = \{\mathbf{p}\}$, so $\mathbf{p}$ is not a limit point of S.

No point $\mathbf{p} \in \mathbb{R}^k$ is a limit point of S, and since $E \subset \mathbb{R}^k$, we see that S has no limit point in E.

For the other case, assume E is not closed. Then it has a limit point $\mathbf{x}_0 \in \mathbb{R}^k$ with $\mathbf{x}_0 \notin E$. We can again construct an infinite subset S of E whose only limit point is $\mathbf{x}_0$, so it has no limit point in E. For every $r > 0$, there is a point $\mathbf{x} \in E$ such that $|\mathbf{x} - \mathbf{x}_0| < r$, so for every $n \in \mathbb{N}$, there is some $\mathbf{x}_n \in E$ with $|\mathbf{x}_n - \mathbf{x}_0| < \frac{1}{n}$. Let $S = \{\mathbf{x}_n \mid n \in \mathbb{N}\}$, so $S \subset E$.

Also, S is infinite. Why? If S had a finite number of points $\{\mathbf{x}_1, \mathbf{x}_2, \ldots, \mathbf{x}_N\}$, then there would exist a j (between 1 and N) where $|\mathbf{x}_j - \mathbf{x}_0|$ is less than $\frac{1}{n}$ for every integer $n \geq N$. This would imply $\mathbf{x}_j = \mathbf{x}_0$ (because otherwise, by the Archimedean property, there would be an n such that $n\,|\mathbf{x}_j - \mathbf{x}_0| > 1$). Then we would have $\mathbf{x}_N \notin E$, which is a contradiction.

Clearly $\mathbf{x}_0$ is a limit point of S. Why? For every $r > 0$, by the Archimedean property there exists an $n \in \mathbb{N}$ such that $nr > 1$. Then $|\mathbf{x}_0 - \mathbf{x}_n| < \frac{1}{n} < r$, so at least one $\mathbf{x}_n$ belongs to $N_r(\mathbf{x}_0) \cap S$.

But for any $\mathbf{y} \in \mathbb{R}^k$ with $\mathbf{y} \neq \mathbf{x}_0$, we have

$$|\mathbf{x}_n - \mathbf{y}| \geq |\mathbf{x}_0 - \mathbf{y}| - |\mathbf{x}_n - \mathbf{x}_0| \quad \text{(by Property 6 of Theorem 6.11)}$$
$$> |\mathbf{x}_0 - \mathbf{y}| - \frac{1}{n}$$
$$\geq \frac{1}{2}|\mathbf{x}_0 - \mathbf{y}| \quad \text{whenever } \frac{1}{n} \leq \frac{1}{2}|\mathbf{x}_0 - \mathbf{y}|.$$

In other words, for $r = \frac{1}{2}|\mathbf{x}_0 - \mathbf{y}|$, every point of $\mathbf{x}_n$ is a distance of at least r away from $\mathbf{y}$ whenever $\frac{1}{n} \leq \frac{1}{2}|\mathbf{x}_0 - \mathbf{y}|$. Thus

$$N_r(\mathbf{y}) \cap S = \left\{ \mathbf{x}_n \mid \frac{1}{n} > \frac{1}{2}|\mathbf{x}_0 - \mathbf{y}| \right\}.$$

There are only finitely many $n \in \mathbb{N}$ with $n < \frac{2}{|\mathbf{x}_0 - \mathbf{y}|}$, so every neighborhood of $\mathbf{y}$ can only contain a finite number of points of S. So by Theorem 9.11, $\mathbf{y}$ cannot be a limit point of S.

Now no point of E is a limit point of S. □

Note that for any $E \subset \mathbb{R}^k$, the following three statements are equivalent:

Statement 1. E is closed and bounded.
Statement 2. E is compact.
Statement 3. Every infinite subset of E has a limit point in E.

It turns out that the second two statements are equivalent in *any* metric space (not just in $\mathbb{R}^k$), but the proof of *Statement 3* $\implies$ *Statement 2* is very complex and unnecessary for our purposes.

It is interesting that we first discovered some properties that all compact sets have (being closed and bounded, and containing the limit point of every infinite subset), and then proved that those properties imply compactness in $\mathbb{R}^k$. Not all properties of compact sets imply such a converse statement, however. For example, the collection of nested sets $\{A_n = (-\frac{1}{n}, \frac{1}{n})\}$ has a non-empty infinite intersection (since $\bigcap_{n=1}^{\infty} A_n = \{0\}$), but each A_n is not compact.

The following theorem is an example of why studying compact sets is helpful. Even though compactness is nowhere to be seen in the statement of the theorem, we use previous compactness theorems to prove a general (and quite useful) result.

Theorem 12.8. *(The Weierstrass Theorem)*
If an infinite subset E of $\mathbb{R}^k$ is bounded, then E has a limit point in $\mathbb{R}^k$.

Proof. Piece of cake. Fill in the blanks in Box 12.1.

BOX 12.1

PROVING THE WEIERSTRASS THEOREM

By Theorem _____, $E \subset I$ for some k-cell I. By Theorem 12.4, I is _____, so by Theorem 11.13, E has a limit point in _____. Well, I is a subset of $\mathbb{R}^k$, so E has a _____ in $\mathbb{R}^k$.

This concludes our fun-filled, rollicking misadventures in the world of compact sets. (Don't worry, though—they'll come back to visit us in Chapter 14!)

In the next chapter, we'll study perfect sets in more detail and learn about connected sets.

CHAPTER 13

Perfect and Connected Sets

While exploring closed, open, and compact sets in detail, we have so far ignored perfect sets. Poor perfect sets, all alone in the corner. Well, this chapter is their chance to shine!

We'll then close our study of topology by introducing the idea of *connected* sets, which, like compact sets, will play a big role in the theory of continuous functions.

Remember we saw in Example 9.25 that any closed interval $[a, b]$ is perfect—since not only does it contain all of its limit points, but in fact every one of its elements is a limit point.

We learned in Example 8.10 that the open interval $(0, 1)$ is uncountable, and using that same Cantor diagonal process, we could have found that any interval $[a, b]$ is uncountable.

It turns out that not only are intervals uncountable, but in fact every real perfect set is uncountable.

Theorem 13.1. *(Real Perfect Sets Are Uncountable)*
If a nonempty subset P of $\mathbb{R}^k$ is perfect, then P is uncountable.

The statement "P is uncountable" really means "$|P|$ is uncountable" (i.e., P has an uncountable number of elements), in the same way we usually write "P is infinite" to mean "$|P|$ is infinite."

Notice that instead of using the Cantor diagonal process, we could have first proved this theorem to show that any interval $[a, b]$ is uncountable (and therefore so is $\mathbb{R}$, since $[a, b] \subset \mathbb{R}$).

Proof. This proof will actually make use of—surprise!—compact sets. Because P is nonempty, it contains at least one point $\mathbf{x}$. And because P is perfect, $\mathbf{x}$ is a limit point, so every neighborhood contains an infinite number of points of P, so P must be infinite. This theorem asserts that P is uncountable; so if we assume P is countable, we can number its points $P = \{\mathbf{x}_1, \mathbf{x}_2, \mathbf{x}_3, \ldots\}$, and we should get a contradiction.

We will make use of Corollary 11.12, so we want a sequence of nested compact sets. We also want to take advantage of the fact that we have numbered P's points, and all of its points are limit points, so let's work with neighborhoods of points in P. Neighborhoods are already bounded, so in order for them to be compact, we want them also to be closed. Thus, instead of making neighborhoods, we'll take neighborhoods whose *closures* are nested.

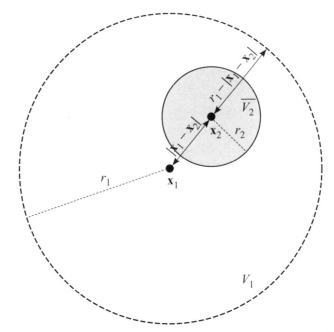

Figure 13.1. Our choice of r_2 guarantees that $\overline{V_2} \subset V_1$ and that $\mathbf{x}_1 \notin \overline{V_2}$.

Choose a point $\mathbf{x}_1 \in P$ and any radius $r_1 > 0$, and let $V_1 = N_{r_1}(\mathbf{x}_1)$. Note that

$$\overline{V_1} = \{\mathbf{y} \in \mathbb{R}^k \mid |\mathbf{x}_1 - \mathbf{y}| \leq r_1\}.$$

Since $\mathbf{x}_1$ is a limit point of P, V_1 must contain at least one other point $\mathbf{x}_2 \in P$ with $\mathbf{x}_2 \neq \mathbf{x}_1$. Now, as seen in Figure 13.1, choose $r_2 > 0$ with

$$r_2 < \min\{|\mathbf{x}_1 - \mathbf{x}_2|,\ r_1 - |\mathbf{x}_1 - \mathbf{x}_2|\},$$

and let $V_2 = N_{r_2}(\mathbf{x}_2)$. Then we have $\overline{V_2} \subset V_1$, and because the distance between $\mathbf{x}_1$ and $\mathbf{x}_2$ is greater than r_2, also $\mathbf{x}_1 \notin \overline{V_2}$. Since $\mathbf{x}_2$ is also a limit point of P, $V_2 \cap P$ must contain at least one other point $\mathbf{x}_3 \in P$ with $\mathbf{x}_3 \neq \mathbf{x}_2$, and since $\mathbf{x}_3 \in V_2$, also $\mathbf{x}_3 \neq \mathbf{x}_1$. We proceed as before to choose a new radius r_3, and continue for an infinite number of steps.

To make this construction more formal, let's start with our definition of V_1. We now need a rule that says how to get to V_{n+1} from V_n. Well, having defined V_n as the neighborhood of a point $\mathbf{x}_n \in P$, we construct V_{n+1} as follows. Since $\mathbf{x}_n$ is a limit point of P, $V_n \cap P$ must contain at least one other point $\mathbf{x}_{n+1} \in P$ with $\mathbf{x}_{n+1} \neq \mathbf{x}_n$. Now choose $r_{n+1} > 0$ with

$$r_{n+1} < \min\{|\mathbf{x}_n - \mathbf{x}_{n+1}|,\ r_n - |\mathbf{x}_n - \mathbf{x}_{n+1}|\},$$

and let $V_{n+1} = N_{r_{n+1}}(\mathbf{x}_{n+1})$. Then we have $\overline{V_{n+1}} \subset V_n$, and since the distance between $\mathbf{x}_n$ and $\mathbf{x}_{n+1}$ is greater than r_{n+1}, also $\mathbf{x}_n \notin \overline{V_{n+1}}$.

Cool! Now let $K_n = \overline{V_n} \cap P$. By the Heine-Borel theorem, $\overline{V_n}$ is compact because it is closed and bounded. And since P is closed, Corollary 11.9 implies that K_n is compact.

Figure 13.2. A computer-generated image of the first five steps in this process. The first row (just a black line) corresponds to E_0, and the black region of each subsequent nth row corresponds to E_{n-1}.

For every $n \in \mathbb{N}$, we have

$$\overline{V_{n+1}} \subset V_n \implies \overline{V_{n+1}} \subset \overline{V_n}$$
$$\implies \left(\overline{V_{n+1}} \cap P\right) \subset \left(\overline{V_n} \cap P\right)$$
$$\implies K_{n+1} \subset K_n.$$

Each V_n is a neighborhood of a limit point of P, so each V_n contains at least one point of P, so each K_n has at least one element. We thus have a sequence of nested, nonempty compact sets $\{K_n\}$, so Corollary 11.12 implies $\bigcap_{n=1}^{\infty} K_n \neq \emptyset$.

Wait a minute! Remember that for any $n \in \mathbb{N}$, $\mathbf{x}_n \notin \overline{V_{n+1}}$, so $\mathbf{x}_n \notin K_{n+1}$, so $\mathbf{x}_n \notin \bigcap_{i=1}^{\infty} K_i$. This is where countability becomes an issue. For any given n, $\mathbf{x}_n \notin \bigcap_{i=1}^{\infty} K_i$, so no element of P is in the intersection, giving us a contradiction. Thus P must be uncountable. □

At this point, you've probably thought to yourself that closed intervals are the simplest example of perfect sets in $\mathbb{R}$. You might have assumed, therefore, that every perfect set in $\mathbb{R}$ is a superset of some interval. But you would be wrong! The infamous *Cantor set* is a great counterexample. It is real and perfect (just like you, you special snowflake), but not only is it *not* a superset of any closed interval—in fact, it doesn't even contain a single open interval.

To construct the Cantor set, start with the set $E_0 = [0, 1]$. Then remove its middle third to make $E_1 = E_0 \setminus (\frac{1}{3}, \frac{2}{3})$, so $E_1 = [0, \frac{1}{3}] \cup [\frac{2}{3}, 1]$. To get E_2, remove the middle third of each of the intervals in E_1, so $E_2 = [0, \frac{1}{9}] \cup [\frac{2}{9}, \frac{3}{9}] \cup [\frac{6}{9}, \frac{7}{9}] \cup [\frac{8}{9}, 1]$. Continue this process for an infinite number of steps (see Box 13.1). (The first five steps are shown in Figure 13.2.)

BOX 13.1

> STEP 3 OF CONSTRUCTING THE CANTOR SET
>
> For example, the set E_3 is the set E_2 without the open intervals _____, _____, _____, and _____.
> Thus $E_3 =$ _____ ∪ _____ ∪ _____ ∪ _____ ∪ _____ ∪ _____ ∪ _____ ∪ _____.

Notice that at every step of the process, we are splitting each closed interval of E_n into two closed intervals (by removing its middle third), so the number of closed intervals in E_n is twice the number of closed intervals in E_{n-1}, which is $2(2^{n-1}) = 2^n$. Also, the size of each closed interval of E_n is one third the size of each closed interval of E_{n-1}, meaning each closed interval of E_n has length $\left(\frac{1}{3}\right)^n$.

The Cantor set is given by the infinite intersection of all the sets E_n. We will formalize this construction in a definition.

First note that when anything—such as a set—is defined recursively (in terms of the previous set), the construction must work something like a proof by induction. We must have a "base case" that specifies what the first set looks like, and then an "inductive step," which says "given that set n looks like this, here is what set $n + 1$ looks like." For the construction to be valid, set $n + 1$ must end up satisfying the same assumptions that set n did, so that set $n + 2$ can be constructed in the same way, and so on.

Definition 13.2. *(The Cantor Set)*
Let $E_0 = [0, 1]$. Then given

$$E_n = [a_0, b_0] \cup [a_1, b_1] \cup \ldots \cup [a_{2^n-1}, b_{2^n-1}],$$

let

$$E_{n+1} = E_n \setminus \left\{ \left(a_0 + \frac{b_0 - a_0}{3}, a_0 + \frac{2(b_0 - a_0)}{3}\right) \cup \right.$$

$$\left(a_1 + \frac{b_0 - a_0}{3}, a_1 + \frac{2(b_0 - a_0)}{3}\right) \cup \ldots$$

$$\left. \left(a_{2^n-1} + \frac{b_0 - a_0}{3}, a_{2^n-1} + \frac{2(b_0 - a_0)}{3}\right) \right\}.$$

The **Cantor set** *is the set*

$$P = \bigcap_{n=0}^{\infty} E_n.$$

We didn't formally prove that E_{n+1} is the union of 2^{n+1} intervals, and thus that the construction is valid. If we wanted to prove it explicitly, it would require a few lines of induction.

Theorem 13.3. *(The Cantor Set Is Nonempty)*
The Cantor set P contains at least one element.

Proof. Notice that each E_n is a finite union of intervals (which are closed sets), so by Theorem 10.4 each E_n is also closed. Of course each E_n is bounded, so by the Heine-Borel theorem, each E_n is compact. They are nested, because each E_{n+1} is built from E_n without some of E_n's elements. Then by Corollary 11.12, $\bigcap_{n=1}^{\infty} E_n \neq \emptyset$, so the Cantor set P contains at least one element. □

Also notice that P itself is closed (since it is an infinite intersection of closed sets), and obviously P is bounded, so P is compact.
As promised, we'll see that P contains no open interval, but is indeed perfect.

Theorem 13.4. *(The Cantor Set Contains No Open Interval)*
For any $a, b \in \mathbb{R}$ with $a < b$, the open interval (a, b) is not a subset of the Cantor set P.

Proof. The closed interval $[a, b]$ has length $b - a$. Remember that for any $n \in \mathbb{N}$, E_n is the union of closed intervals, each of length 3^{-n}. Because $P = \bigcap_{n=1}^{\infty} E_n$, P is a subset of every E_n. So given any $n \in \mathbb{N}$, P contains no closed interval—and thus no open interval—of length greater than 3^{-n}.

By the Archimedean property of $\mathbb{R}$, we can choose an $n \in \mathbb{N}$ such that

$$n > -\frac{\log(b - a)}{\log(3)}.$$

Then $\log(b - a) > -n \log(3)$, so $b - a > 3^{-n}$. Thus P cannot contain (a, b). □

 Isn't this sort of cheap? I mean, we constructed the Cantor set by building unions of closed intervals, and now we used the fact that these closed intervals get smaller and smaller to show that the Cantor set cannot contain any open interval (or closed interval). How can there be no open intervals (or closed intervals) if the Cantor set is supposedly a union of closed intervals?

Note the key problem with this thinking: the Cantor set itself is *not* a union of closed intervals; it is an infinite intersection of unions of closed intervals. As we've seen before, infinite intersections of closed intervals could contain as little as a single point, as with $\bigcap_{n=1}^{\infty} \left[-\frac{1}{n}, \frac{1}{n}\right] = \{0\}$.

Theorem 13.5. *(The Cantor Set Is Perfect)*
The Cantor Set P is perfect in the metric space $\mathbb{R}$.

Proof. We already know P is closed in $\mathbb{R}$, so we just need to show that every point of P is a limit point of P. Take any $x \in P$, and for any $r > 0$, let

$$S = N_r(x) = (x - r, x + r).$$

We want to show that some other point of P is in the open interval S.

Well, $x \in \bigcap_{n=1}^{\infty} E_n$, so for every $n \in \mathbb{N}$, $x \in E_n$. We know E_n is a union of intervals, so x must be in some interval $I_n \subset E_n$.

The length of the open interval S is $2r$, and by Definition 13.2, the length of the interval I_n is 3^{-n}. By the Archimedean property of $\mathbb{R}$, we can choose an $n \in \mathbb{N}$ such that

$$n > -\frac{\log(2r)}{\log(3)}.$$

Thus $3^{-n} < 2r$, so $I_n \subset S$. Take an endpoint x_n of I_n such that $x_n \neq x$ (meaning, if $I_n = [a, x]$ take $x_n = a$; if $I_n = [x, b]$ take $x_n = b$; if $I_n = [a, b]$ take $x_n = a$ or $x_n = b$). Then $x_n \in I_n$, so $x_n \in S$, so we have shown that some non-x point is in every neighborhood of x.

Thus x is a limit point of P, and since x was arbitrary, P is perfect. □

Note that by Theorem 13.1, this theorem also implies that the Cantor set is uncountable.

We now shift our attention to connected sets, which can be a little tricky. To better understand them, we'll present two different definitions and then prove they are equivalent. (In other words, we'll *connect* them!) Then, we'll prove a theorem that gives a more intuitive characterization of connected sets on the real line.

We'll start by defining *separated* sets, which are the basis for connectedness.

Definition 13.6. *(Separated Sets)*
Two subsets A and B of a metric space X are **separated** if A does not intersect the closure of B and B does not intersect the closure of A.

In symbols, $A, B \subset X$ are separated if:

$$A \cap \overline{B} = \emptyset \text{ and } \overline{A} \cap B = \emptyset.$$

Example 13.7. *(Separated Sets)*
Remember that the point $a \in \mathbb{R}$ is a limit point of the half-open interval $(a, b]$. Thus the sets $A = [-3, 0)$ and $B = (0, 3]$ are separated, since $\overline{A} = [-3, 0]$ does not intersect $B = (0, 3]$, and $\overline{B} = [0, 3]$ does not intersect $A = [-3, 0)$.

As a further example, for any $n \in \mathbb{N}$, every pair of closed intervals in the set E_n (from our construction of the Cantor set) are separated.

Remember from Definition 3.10 that two sets are *disjoint* if they do not intersect. For example, $A = [-3, 0]$ and $B = (0, 3]$ are disjoint. However, A and B are not separated, since $A \cap \overline{B} = [-3, 0] \cap [0, 3] = \{0\} \neq \emptyset$. It is helpful to think of separatedness as a stronger version of disjointedness.

This list should make the distinction more explicit:

$[a, b)$ and $(b, c]$	disjoint, separated
$[a, b)$ and $[b, c]$	disjoint, not separated
$[a, b]$ and $(b, c]$	disjoint, not separated
$[a, b]$ and $[b, c]$	not disjoint, not separated

Of course, the same properties hold if we replace $[a$ with $(a$ and/or if we replace $c]$ with $c)$.

Theorem 13.8. *(Separated Subsets)*
For any two subsets A and B of a metric space X, let $A_1 \subset A$ and let $B_1 \subset B$. If A and B are separated, then A_1 and B_1 are also separated.

Proof. We know $A \cap \overline{B} = \overline{A} \cap B = \emptyset$, and we want to show $A_1 \cap \overline{B_1} = \overline{A_1} \cap B_1 = \emptyset$.

Notice that $\overline{A_1} \subset \overline{A}$ (and similarly $\overline{B_1} \subset \overline{B}$). This is actually a general property of closures. Why does it work? Take $a_1 \in \overline{A_1}$. If $a_1 \in A_1$, then $a_1 \in A$ (since $A_1 \subset A$), so $a_1 \in \overline{A}$ (since $A \subset \overline{A}$). If $a_1 \notin A_1$, then a_1 is a limit point of A_1, so every neighborhood of a_1 intersects at least one other point of A_1. But every point of A_1 is a point of A, so every neighborhood of a_1 intersects at least one other point of A. Thus a_1 is a limit point of A, so $a_1 \in \overline{A}$.

Now we have

$$\left(A_1 \cap \overline{B_1}\right) \subset \left(A \cap \overline{B_1}\right) \subset \left(A \cap \overline{B}\right) = \emptyset,$$

and similarly

$$\left(\overline{A_1} \cap B_1\right) \subset \left(\overline{A_1} \cap B\right) \subset \left(\overline{A} \cap B\right) = \emptyset.$$

□

Definition 13.9. *(Connected Set)*
A subset E of a metric space X is **disconnected** if it is the union of two nonempty separated sets.

In symbols, $E \subset X$ is disconnected if $\exists A, B \subset X$ such that:

1. $A \neq \emptyset$ and $B \neq \emptyset$.
2. $E = A \cup B$.
3. $A \cap \overline{B} = \emptyset$ and $\overline{A} \cap B = \emptyset$.

*A subset E of a metric space X is **connected** if it is not disconnected.*

Example 13.10. (Connected Sets)
Since the definition of disconnectedness states that the separated sets A and B must be nonempty, this implies that the empty set $\emptyset$ is connected.

The set $E = [-3, 0) \cup (0, 3]$ is disconnected, since we saw in Example 13.7 that $[-3, 0)$ and $(0, 3]$ are separated.

On the other hand, even though $[-3, 0]$ and $(0, 3]$ are not separated, we cannot say for sure that $E = [-3, 0] \cup (0, 3] = [-3, 3]$ is connected—because how do we know there are not two other sets A and B with $A \cup B = [-3, 3]$ that *are* separated? (It turns out that all open and closed intervals are in fact connected, but this requires an actual proof, which we will do later.)

As a further example, the Cantor set P is disconnected. Why? Let A be the points of P that are $\leq \frac{1}{2}$, so $A = P \cap (-\infty, \frac{1}{2}]$; and let B be the points of P that are $> \frac{1}{2}$, so $B = P \cap (\frac{1}{2}, \infty)$. Then $P = A \cup B$, and both A and B are nonempty. Since $P \subset E_1$, we have $A \subset [0, \frac{1}{3}]$ and $B \subset [\frac{2}{3}, 1]$. Since $[0, \frac{1}{3}]$ and $[\frac{2}{3}, 1]$ are separated, so are A and B by Theorem 13.8. Thus P is the union of two nonempty separated sets.

There is an alternate definition for disconnectedness, which you might find to be easier to work with in some cases. In this theorem, we will prove that the two definitions are equivalent.

Theorem 13.11. *(Alternate Definition of Disconnectedness)*
A subset E of a metric space X is disconnected if and only if $\exists U, V \subset X$ such that

1. U and V are open.
2. $E \subset U \cup V$.
3. $E \cap U \neq \emptyset$ and $E \cap V \neq \emptyset$.
4. $E \cap U \cap V = \emptyset$.

Proof. We'll start with the first direction of the equivalence, in which we assume E is disconnected according to Definition 13.9, and prove it satisfies the conditions in this theorem. We thus assume that there are nonempty sets A and B with $E = A \cup B$, $A \cap \overline{B} = \emptyset$, and $\overline{A} \cap B = \emptyset$.

Let $U = \overline{A}^C$ (the complement of $\overline{A}$ in X), and let $V = \overline{B}^C$ (the complement of $\overline{B}$ in X).

1. Since $\overline{A}$ and $\overline{B}$ are closed (relative to X) by Theorem 10.7, U and V are open (relative to X) by Theorem 10.1.
2. Since $A \cap \overline{B} = \emptyset$, we have $A \subset \overline{B}^C$; since $\overline{A} \cap B = \emptyset$, we have $B \subset \overline{A}^C$. Thus

$$E = (A \cup B) \subset \left(\overline{A}^C \cup \overline{B}^C\right) = U \cup V.$$

3. Since $B \subset E$ and $B \neq \emptyset$, we have

$$E \cap U = (E \cap \overline{A}^C) \supset (E \cap B) = B \neq \emptyset,$$

and since $A \subset E$ and $A \neq \emptyset$, we have

$$E \cap V = (E \cap \overline{B}^C) \supset (E \cap A) = A \neq \emptyset.$$

4. We can calculate

$$E \cap U \cap V = E \cap (\overline{A}^C \cap \overline{B}^C)$$
$$= E \cap (\overline{A} \cup \overline{B})^C \quad \text{(by De Morgan's law)}$$
$$\subset E \cap E^C \quad \text{(since } E \subset \overline{A} \cup \overline{B}, \text{ so } E^C \supset (\overline{A} \cup \overline{B})^C\text{)}$$
$$= \emptyset.$$

To prove the other direction of the equivalence, we assume E is disconnected according to the conditions of this theorem and prove it satisfies Definition 13.9. We thus assume that there are open sets U and V with $E \subset U \cup V$, $E \cap U \neq \emptyset$, and $E \cap V \neq \emptyset$, but $E \cap U \cap V = \emptyset$.

Let $A = E \cap U$ and let $B = E \cap V$. Proving the properties of connectedness is just a matter of using logic and set theory. It's always helpful to practice this stuff, so try filling in the blanks in Box 13.2.

BOX 13.2

PROVING THE OTHER DIRECTION OF THEOREM 13.11
1. $A = E \cap U \neq$ _____ and $B =$ _____ $\neq \emptyset$.
2. $A \cup B = (E \cap U) \cup (E \cap V) = E \cap$ _____ $= E$, since $E \subset U \cup V$.
3. $A \cap \overline{B} = (E \cap U) \cap \overline{(E \cap V)} \subset (E \cap U) \cap$ _____ $= E \cap V \cap U = \emptyset$, and $\overline{A} \cap B =$ _____ $\cap (E \cap V) \subset (E \cap U) \cap$ _____ $= E \cap V \cap U =$ ____.

□

Theorem 13.12. *(Connected Sets on the Real Line)*
A subset E of $\mathbb{R}$ is connected if and only if given any two points $x, y \in E$ and given $z \in \mathbb{R}$ such that $x < z < y$, then $z \in E$.

Proof. We need to show both directions of the theorem:

$$\text{connected} \implies \text{includes all middle points},$$
$$\text{includes all middle points} \implies \text{connected}.$$

It will be easier to prove the contrapositives of each direction; namely,

excludes at least one middle point $\implies$ *disconnected*,

disconnected $\implies$ *excludes at least one middle point*.

For this proof, it happens to be easier to use Definition 13.9 rather than Theorem 13.11.

We first assume there exist $x, y \in E$ and $z \in \mathbb{R}$ with $x < z < y$, but $z \notin E$. Let A be the set of all numbers in E that are less than z, so $A = E \cap (-\infty, z)$; and let B be the set of all numbers in E which are greater than z, so $B = E \cap (z, \infty)$. Then $E = A \cup B$, and both A and B are nonempty (since $x \in A$ and $y \in B$). Notice that $A \subset (-\infty, z)$ and $B \subset (z, \infty)$, and since $(-\infty, z)$ and (z, ∞) are separated, by Theorem 13.8 so are A and B. Thus E is the union of two nonempty separated sets.

Now we assume E is disconnected, so there exist nonempty separated sets A and B such that $E = A \cup B$. There must be at least one element $x \in A$, and at least one element $y \in B$. We know $x \neq y$ (otherwise $A \cap B \neq \emptyset$), so we can assume without loss of generality that $x < y$. (If $x > y$, the same proof works—we just switch A and B.) Let

$$z = \sup\{a \in A \mid x \leq a \leq y\} = \sup(A \cap [x, y]).$$

We want to prove that this z is not an element of E. Basically, we will take advantage of the separatedness of A and B to show that there is "stuff" in between those two sets (so that stuff is outside E). It shouldn't be too hard to prove that $z \notin B$. Then, if $z \notin A$, we're good—otherwise, we just take an element z_1 that is close enough to z to not be in A or B.

To make this argument formal, we note that $A \cap [x, y]$ is bounded above (by y), so by Theorem 10.10, $z \in \overline{A \cap [x, y]} = \overline{A} \cap [x, y]$. Then $z \in \overline{A}$, so z cannot be an element of B (otherwise $\overline{A} \cap B \neq \emptyset$). We know $x \leq z \leq y$, and $z \neq y$ (since $z \notin B$), so $x \leq z < y$.

We now have two possible cases.

Case 1. If $z \notin A$, then $z \neq x$, so we have $x < z < y$. Also, $z \notin B$, so $z \notin E$, which is exactly what we wanted to prove.

Case 2. If $z \in A$, then $z \notin \overline{B}$ (otherwise $A \cap \overline{B} \neq \emptyset$). Because z is not a limit point of B, there must be some neighborhood of z that does not intersect B, meaning there exists an $r > 0$ such that $(z - r, z + r) \cap B = \emptyset$. Pick $z_1 \in (z, z + r)$, so that $z < z_1 < y$ (note that $z_1 \neq y$ since $z_1 \notin B$), and we have $x < z_1 < y$. Also, z_1 is strictly greater than any element of $A \cap [x, y]$ (since $z_1 > z = \sup A \cap [x, y]$), so it cannot be in $A \cap [x, y]$. Of course $z_1 \in [x, y]$, meaning z_1 cannot be an element of A. Thus $z_1 \notin A$ and $z_1 \notin B$, so $z_1 \notin E$.

$\square$

Corollary 13.13. *(Open and Closed Intervals Are Connected)*
Every open interval (a, b) and every closed interval $[a, b]$ is connected.

Proof. Open and closed intervals are subsets of $\mathbb{R}$. By definition, they include all real numbers between a and b. So given x and y in (a, b) or $[a, b]$, any $z \in \mathbb{R}$ with $x < z < y$ is included in the interval. Thus by Theorem 13.12, they are connected. $\square$

This chapter concludes our study of topology. You've survived a barrage of new definitions, and hopefully along the way you've learned some techniques for dealing with them. In the future, whenever you encounter a problem that uses a new definition, you should first try to fully internalize the definition before applying it to the problem. Write out what it means in both words and symbols, play with some basic examples, understand how it works in $\mathbb{R}$ or $\mathbb{R}^k$, and draw pictures.

You'll notice that we had a good mix of material in generic metric spaces X and in $\mathbb{R}$ or $\mathbb{R}^k$ specifically. Although real analysis is primarily concerned with real numbers, the topology we learned for generic metric spaces will prove just as useful.

Coming up next: sequences! So exciting!

SEQUENCES

CHAPTER 14

Convergence

We'll begin our study of sequences and series by exploring the notion of *convergence*, which basically asks, "as a sequence continues to infinity, does it get arbitrarily close to some point?"

Though we introduced sequences in Chapter 2, we should first give a more formal definition, which is based on what we learned about functions in Definition 8.3.

Definition 14.1. *(Sequence)*
A **sequence** in a metric space X is a function

$$f: \mathbb{N} \to X, \quad f: n \mapsto s_n$$

where $p_n \in X, \forall n \in \mathbb{N}$. We denote the sequence as $\{p_n\}$ or as $p_1, p_2, \ldots$

The set of all possible values of the sequence is called the **range** of $\{p_n\}$. The sequence is **bounded** if its range is bounded in X (according to Definition 9.3).

Example 14.2. *(Sequences)*
By this definition, every countable set can be made into a sequence. In particular, $\mathbb{Q}$ can be arranged into a sequence, but $\mathbb{R}$ cannot.

Remember that all sequences go on for an infinite number of steps, but the range of a sequence might be finite if the sequence repeats. For example, the set $\{1\}$ is not a sequence in and of itself, but we can make it into a sequence by writing $1, 1, 1, \ldots$

Let's look at the following sequences in the metric space $\mathbb{R}$:

1. If $s_n = \frac{1}{n}$ for every $n \in \mathbb{N}$, then the range of $\{s_n\}$ is the set $\{1, \frac{1}{2}, \frac{1}{3}, \ldots\}$. This range is infinite and bounded (since the distance between any two elements is ≤ 1). Thus we say that $\{s_n\}$ is bounded.
2. If $s_n = n^2$ for every $n \in \mathbb{N}$, then the range of $\{s_n\}$ is the set $\{1, 4, 9, \ldots\}$. This range is infinite and unbounded (since the numbers get larger and larger). Thus we say that $\{s_n\}$ is unbounded.
3. If $s_n = 1 + \frac{(-1)^n}{n}$ for every $n \in \mathbb{N}$, then the range of $\{s_n\}$ is the set $\{0, \frac{2}{3}, \frac{4}{5}, \frac{6}{7}, \ldots\} \cup \{\frac{3}{2}, \frac{5}{4}, \frac{7}{6}, \ldots\}$. This range is infinite and bounded (since the distance between any two elements is $\leq \frac{3}{2}$). Thus we say that $\{s_n\}$ is bounded.

4. If $s_n = 1$ for every $n \in \mathbb{N}$, then the range of $\{s_n\}$ is the set $\{1\}$. This range is finite and bounded (because the distance between any two elements is ≤ 0). Thus we say that $\{s_n\}$ is bounded.
5. For this example, we'll use the metric space $\mathbb{C}$ instead of $\mathbb{R}$. If $s_n = i^n$ for every $n \in \mathbb{N}$ (where $i^2 = -1$ as in Chapter 6), then the range of $\{s_n\}$ is the set $\{i, -1, -i, 1\}$. This range is finite and bounded (since the distance between any two elements is ≤ 2). Thus we say that $\{s_n\}$ is bounded.

Definition 14.3. *(Convergence)*
For any sequence $\{p_n\}$ in a metric space X, $\{p_n\}$ **converges** to the point $p \in X$ if for every $\epsilon > 0$, $d(p_n, p) < \epsilon$ for every n greater than or equal to some natural number N. We say p is the **limit** of $\{p_n\}$.
  In symbols, we write $\lim_{n \to \infty} p_n = p$ (or $p_n \to p$ for short) if:

$$\forall \epsilon > 0, \exists N \in \mathbb{N} \text{ such that } n \geq N \implies d(p_n, p) < \epsilon.$$

A sequence $\{p_n\}$ **diverges** if it does not converge to any $p \in X$.

Distinction. A limit is *not* a limit point. The former relates to sequences, while the latter relates to topology. These ideas *are* connected, though, as we will see later on in this chapter.

What's all this business with with the $\forall \epsilon$ and $\exists N$? For $\{p_n\}$ to converge to p, we need to be sure that the following is true: given any arbitrarily small distance ϵ, there is some step N past which every element of the sequence is closer than the distance of ϵ to its limit p. (Sometimes you'll see N_ϵ, to emphasize that this step number N depends on the distance ϵ.)

For example, if the sequence $\{p_n\}$ in $\mathbb{R}$ converges to the number 1, there is some number N such that $p_N, p_{N+1}, p_{N+2}, \ldots$ are all between 0.9 and 1.1. Similarly, there is another number N such that $p_N, p_{N+1}, p_{N+2}, \ldots$ are all between 0.95 and 1.05, and so on. Because this is true for *every* possible distance ϵ, we are sure the sequence gets infinitely close to the point 1 (even though it may not ever actually "touch" it).

The N_ϵ Challenge. Another way to think of this is as follows: to prove that a sequence converges to p, you need to pass the N_ϵ challenge. Your friend comes to you and says "ϵ is 0.1." You must respond by finding a natural number N such that $p_N, p_{N+1}, p_{N+2}, \ldots$ are all a distance of less than 0.1 away from p. Then your friend says, "Fine, that one was easy. Now let $\epsilon = 0.00456$, ha!" You need to keep your cool and find another natural number N such that $p_N, p_{N+1}, p_{N+2}, \ldots$ are all a distance of less than 0.00456 away from p. Your friend keeps challenging you with different values of $\epsilon > 0$, over and over again, and you keep parrying by finding an N that works.

Eventually your friend, ever the wily trickster, writes a computer program to keep on dishing out random values of ϵ: $3, 0.241, 100, \frac{\sqrt{\pi}}{7}$, and so on. There are an infinite number of them, and your manual N-finding cannot keep up with the computer. Instead, you must fight fire with fire.

You decide to write your own program, which will take any $\epsilon > 0$, and automatically find a number N such that $p_N, p_{N+1}, p_{N+2}, \ldots$ are all a distance of less than ϵ away from p. To do this, you must provide instructions for how to find an N that works with any possible ϵ. You tell it how to find N_ϵ: a natural number that is a function of the inputted ϵ. For example, $N = \lceil 5 + \frac{3}{\epsilon^2} \rceil$ might work for some sequence.

If this is possible—meaning, if it is possible to have a rule for finding an N for any given $\epsilon > 0$—then you have passed the N_ϵ challenge. Your friend, humbled by defeat, promises to never bother you again (and to not forget your birthday *again* next year). You smile euphorically, because nothing makes you smile like proving that a sequence converges.

Example 14.4. (Convergence)
Let's look at the sequences from the previous example, again in the metric space $\mathbb{R}$:

1. If $s_n = \frac{1}{n}$ for every $n \in \mathbb{N}$, then $\{s_n\}$ converges to 0.
   To prove this, we need to explicitly find an N for each $\epsilon > 0$ such that $d(s_n, 0) < \epsilon$ for every $n \geq N$. Notice that $d(s_n, 0) = |s_n - 0| = \left|\frac{1}{n}\right|$, so as long as $n\epsilon > 1$, we're good. We want $n > \frac{1}{\epsilon}$, so for example, let $N = \lceil\frac{1}{\epsilon}\rceil + 1$. The ceiling function is there to guarantee that N is a natural number, and the $+1$ is there to guarantee that $n\epsilon > 1$, rather than $n\epsilon \geq 1$.
   The whole $N = \lceil\frac{1}{\epsilon}\rceil + 1$ business looks a lot more complicated than it really is. What we are saying is that if $\epsilon = \frac{1}{2}$, take $N = 3$, since $\frac{1}{3}, \frac{1}{4}, \frac{1}{5}, \ldots$ are all less than $\frac{1}{2}$; if $\epsilon = \frac{1}{100.5}$, take $N = 102$, since $\frac{1}{102}, \frac{1}{103}, \frac{1}{104}, \ldots$ are all less than $\frac{1}{100.5}$, and so on.
   Note that we could also choose $N = \lceil\frac{1}{\epsilon}\rceil + 2$, or $N = \lceil\frac{1}{\epsilon}\rceil + 3$, and so on. As long as N is a natural number that is greater than $\frac{1}{\epsilon}$, it will work.
   We can state this proof formally as follows. For every $\epsilon > 0$, choose $N = \lceil\frac{1}{\epsilon}\rceil + 1$. Then for every $n \geq N$, we have

$$d(s_n, 0) = \left|\frac{1}{n}\right| \leq \left|\frac{1}{N}\right|$$

$$= \left|\frac{1}{\lceil\frac{1}{\epsilon}\rceil + 1}\right| < \left|\frac{1}{\frac{1}{\epsilon}}\right| = |\epsilon| = \epsilon.$$

2. If $s_n = n^2$ for every $n \in \mathbb{N}$, then $\{s_n\}$ diverges.
   To prove this, we need to falsely assume that $s_n \to p$ for some real number p and then find a contradiction. If $s_n \to p$, then for every $\epsilon > 0$, there is a natural N such that $n \geq N \implies d(s_n, p) < \epsilon$. But then for every $n \geq N$, $|n^2| - |p| \leq |n^2 - p| < \epsilon$, so $|n^2| < |p| + \epsilon$. This means that the absolute value of p (plus ϵ) would have to be greater than the square of every natural number, which is impossible (unless p were infinite, but infinity is not a valid limit). The contradiction shows that there is no $p \in \mathbb{R}$ with $s_n \to p$.

3. If $s_n = 1 + \frac{(-1)^n}{n}$ for every $n \in \mathbb{N}$, then $\{s_n\}$ converges to 1 (see Figures 14.1 and 14.2).
   To prove this, we need to explicitly find an N for each $\epsilon > 0$ such that $d(s_n, 1) < \epsilon$ for every $n \geq N$. Well,

$$d(s_n, 1) = |s_n - 1| = \left|\frac{(-1)^n}{n}\right| = \left|\frac{1}{n}\right|.$$

Then we choose $N = \lceil\frac{1}{\epsilon}\rceil + 1$ (as in the $s_n = \frac{1}{n}$ example), and we're good.

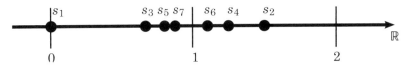

Figure 14.1. The first few elements of the sequence $s_n = 1 + \frac{(-1)^n}{n}$.

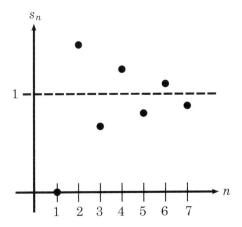

Figure 14.2. The first few elements of the sequence $s_n = 1 + \frac{(-1)^n}{n}$, this time plotted against n.

Note that even though this sequence jumps to the left and right of its limit, as you can see in Figures 14.1 and 14.2, the sequence still converges—since the points to the left and right of 1 both approach 1.

4. If $s_n = 1$ for every $n \in \mathbb{N}$, then $\{s_n\}$ converges to 1.

   To prove this, we need to explicitly find an N for each $\epsilon > 0$ such that $d(s_n, 0) < \epsilon$ for every $n \geq N$. Well, $d(s_n, 1) = |1 - 1| = 0 < \epsilon$ for every $\epsilon > 0$, so $N = 1$ works (or any $N \in \mathbb{N}$, for that matter).

5. Going back to $\mathbb{C}$ for this example, if $s_n = i^n$ for every $n \in \mathbb{N}$, then $\{s_n\}$ diverges.

   This sequence does not converge because, as you can see in Figure 14.3, each of its elements is further than $\epsilon = 1$ from each of its other elements.

   To prove this formally: if $\{s_n\}$ had a limit p, then for every $\epsilon > 0$, there would exist an $N \in \mathbb{N}$ with $d(s_n, p) < \epsilon$ for every $n \geq N$. Since it is true for every $\epsilon > 0$, then given such an $\epsilon > 0$ it is also true for $\frac{\epsilon}{2}$, so we can get

$$d(s_n, s_{n+1}) \leq d(s_n, p) + d(p, s_{n+1}) < \frac{\epsilon}{2} + \frac{\epsilon}{2} = \epsilon.$$

(This trick with $\frac{\epsilon}{2}$ is used pretty often.)

However, because this sequence repeats, we know that for every $N \in \mathbb{N}$, there is some $n \geq N$ with $s_n = i$. Then for $\epsilon = 1$

$$d(s_n, s_{n+1}) = d(i, 1) > 1 = \epsilon,$$

which is a contradiction. Thus s_n cannot have such a limit p.

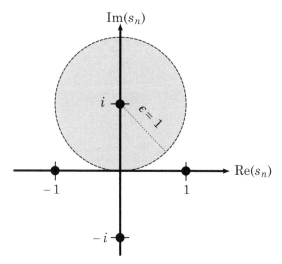

Figure 14.3. All four elements of the sequence $s_n = i^n$. There are no other elements of s_n within a distance of $\epsilon = 1$ from i.

Note that in the metric space $X = \mathbb{R}$, we know the sequence $s_n = \frac{1}{n}$ converges to 0. But if we consider instead s_n as a subset of the metric space $X = \mathbb{R} \setminus \{0\}$, then s_n does not converge to any point of X. Therefore, it is more precise to always write "s_n converges in X" instead of only "s_n converges" (but usually we are lazy and just write "s_n converges" anyway).

The range of a sequence $\{p_n\}$ is a set—in fact, it is a subset of $\{p_n\}$'s metric space X. In Chapter 9, we learned about the limit points of subsets of a metric space. In this chapter, we learned about the limit of a sequence. This leads us to ask the following natural question: what is the difference between the limit of a sequence $\{p_n\}$, and a limit point of the range of $\{p_n\}$? (Recall from Definition 9.9 that a limit point of a set is a point whose every neighborhood contains at least one other point of the set.) It turns out that the two are similar but *not* equivalent, as the following examples will demonstrate.

First, $p_n \to p$ does not imply p *is a limit point of the range of* $\{p_n\}$. To see this, let $p_n = 1$ for every $n \in \mathbb{N}$, so $\{p_n\}$ is the sequence $1, 1, 1, \ldots$, whose range is the set $\{1\}$. This sequence converges, as we saw in the previous example. But $p = 1$ is *not* a limit point of the set $\{1\}$, since every neighborhood around p only contains one point from the set $\{1\}$, which is itself.

Also, p *is a limit point of* $\{p_n\}$ does not imply $p_n \to p$. To see this, let $p_n = (-1)^n + \frac{(-1)^n}{n}$ for every $n \in \mathbb{N}$, so $\{p_n\}$ is the sequence $-2, \frac{3}{2}, -\frac{4}{3}, \frac{5}{4}, -\frac{6}{5}, \frac{7}{6}, \ldots$, whose range is the set $\{-2, -\frac{4}{3}, -\frac{6}{5}, \ldots\} \cup \{\frac{3}{2}, \frac{5}{4}, \frac{7}{6}, \ldots\}$. As we can see in Figures 14.4 and 14.5, this sequence does *not* converge. (Why? For any $n \in \mathbb{N}$, $|p_n - p_{n+1}| > 2$, so we can never have an N with $|p - p_n| < \epsilon$ for every $n \geq N$.) But $p = 1$ *is* a limit point of the range of $\{p_n\}$, since for every $r > 0$, we use the Archimedean property to find an $n \in \mathbb{N}$ with $nr > 1$. Then $1 - r < 1 + \frac{1}{n} < 1 + r$, and so the point $1 + \frac{1}{n} \in N_r(1)$. (We need n to be even here, so if our Archimedean n with $nr > 1$ is odd, just choose $n + 1$.) Similarly, $p = -1$ (with n odd) is also a limit point of the range of $\{p_n\}$.

This demonstrates that the limit of a sequence is different from a limit point of its range. They still look similar—and as the next theorem shows, there is an alternate definition for the limit of a sequence (which more closely resembles the property of limit points that we saw in Theorem 9.11).

Figure 14.4. The first few elements of the sequence $p_n = (-1)^n + \frac{(-1)^n}{n}$.

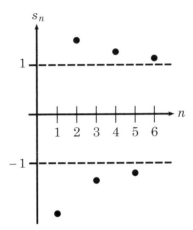

Figure 14.5. The first few elements of the sequence $p_n = (-1)^n + \frac{(-1)^n}{n}$, this time plotted against n.

Theorem 14.5. *(Alternate Definition of Convergence)*
For any sequence $\{p_n\}$ in a metric space X, $\{p_n\}$ converges to $p \in X$ if and only if for every neighborhood of p, there are only finitely many elements of $\{p_n\}$ outside that neighborhood.

Proof. Assume $p_n \to p$. Then for every $\epsilon > 0$, there is some $N \in \mathbb{N}$ such that $d(p_n, p) < \epsilon$ whenever $n \geq N$. In other words, every p_n for $n \geq N$ is a distance of less than ϵ away from p, so $p_n \in N_\epsilon(p)$ for every $n \geq N$. Then $p_n \notin N_\epsilon(p)$ implies $n < N$. Since N is a fixed finite number, there are only finitely many natural numbers less than N, so there are only finitely many elements of $\{p_n\}$ not in $N_\epsilon(p)$. Since this is true for every $\epsilon > 0$, it is true for every neighborhood of p.

Conversely, assume every neighborhood of p contains all of $\{p_n\}$ except for a finite number of its elements. So given $\epsilon > 0$, there is a finite set $\{p_i\}$ of elements of $\{p_n\}$ which are not in $N_\epsilon(p)$. Because $\{p_i\}$ is finite, we can take its greatest index N. Then for every $n \geq N + 1$, $p_n \in N_\epsilon(p)$, so $d(p_n, p) < \epsilon$. Since this is true for every $\epsilon > 0$, we have $p_n \to p$. □

The next theorem proves that if a sequence converges, then its limit is unique.

Theorem 14.6. *(Limits Are Unique)*
For any sequence $\{p_n\}$ in a metric space X, if p_n converges to both $p \in X$ and $p' \in X$, then $p' = p$.

Proof. Here we'll use an argument that you'll see over and over again in real analysis. The basic idea is that if p_n gets arbitrarily close to both p and p', there is some step past which p_n is arbitrarily close to both of them. In that case, p and p' must also be arbitrarily close to each other.

For any $\epsilon > 0$, we can apply the definition of convergence to $\frac{\epsilon}{2}$ to obtain two natural numbers N and N' such that

$$n \geq N \implies d(p_n, p) < \tfrac{\epsilon}{2},$$

$$n \geq N' \implies d(p_n, p') < \tfrac{\epsilon}{2}.$$

Note that N is not necessarily the same as N', so we cannot guarantee, for example, that $n \geq N' \implies d(p_n, p) < \tfrac{\epsilon}{2}$. We can, however, just take the maximum of the two. So for $n \geq \max\{N, N'\}$, we have $n \geq N$ and $n \geq N'$, so

$$d(p, p') \leq d(p, p_n) + d(p_n, p') < \frac{\epsilon}{2} + \frac{\epsilon}{2} = \epsilon.$$

(Notice that we chose $\frac{\epsilon}{2}$ as the arbitrarily small distance for both limits, so that in this step we end up with exactly ϵ. If, for example, we were showing that three limits are unique, we would have wanted to use $\frac{\epsilon}{3}$ for each, so we could end up with $\frac{\epsilon}{3} + \frac{\epsilon}{3} + \frac{\epsilon}{3} = \epsilon$.)

Because this is true for every $\epsilon > 0$, it follows that $d(p, p') = 0$. (See Example 2.2 for the proof of this statement for real numbers; the same argument works for any metric space X.) Then by Definition 9.1, $p' = p$. □

You might have noticed in Example 14.4 that every convergent sequence we looked at was bounded. This turns out to be a general fact, which we will prove in the next theorem.

The contrapositive should be particularly intuitive. If a sequence (such as $s_n = n^2$) has an unbounded range, then surely it does not converge.

Note that the converse, however, is not true. As we saw before, the sequence $s_n = i^n$ is bounded but not convergent.

Theorem 14.7. *(Convergent $\implies$ Bounded)*
For any sequence $\{p_n\}$ in a metric space X, if $\{p_n\}$ converges, then it is bounded.

Proof. To satisfy Definition 9.3, we need to find some point $q \in X$ and some number $M \in \mathbb{R}$ such that $d(p_n, q) \leq M$ for every $n \in \mathbb{N}$. Because $\{p_n\}$ converges, it gets arbitrarily close to its limit $p \in X$. Then the distance between p and any p_n is small, so it is natural to try using $q = p$ in our proof.

Letting $\epsilon = 1$, there is some $N \in \mathbb{N}$ such that $d(p_n, p) < 1$ for every $n \geq N$. All the elements of $\{p_n\}$ for $n < N$ are some distance away from p, and because there are only finitely many of them, we can take the maximum distance. Let

$$r = \max \left\{ d(p_1, p), d(p_2, p), \ldots, d(p_{N-1}, p), 1 \right\}.$$

Then for any $n \in \mathbb{N}$, $d(p_n, p) \leq r$, so $\{p_n\}$ is bounded. □

The following theorem concretely relates limit points of sets to limits of sequences. We can then apply everything we learned about limit points in the previous few chapters to get some interesting corollaries.

Theorem 14.8. *(Converging to Limit Points)*
If $E \subset X$ and if p is a limit point of E, then there is a sequence $\{p_n\}$ in E that converges to p.

Note that the limit point p may or may not be in E. If $p \notin E$, we would say $\{p_n\}$ converges in X, but not in E (since its limit p is in X but not in E).

Proof. Every neighborhood of p contains at least one element of E. So for every $n \in \mathbb{N}$, there is some point p_n of E with $p_n \in N_{\frac{1}{n}}(p)$, meaning $d(p_n, p) < \frac{1}{n}$. We define the sequence $\{p_n\}$ in this manner.

For any $\epsilon > 0$, let $N = \lceil \frac{1}{\epsilon} \rceil + 1$. Then for every $n \geq N$, we have

$$d(p_n, p) < \frac{1}{n} \leq \frac{1}{N}$$

$$= \frac{1}{\lceil \frac{1}{\epsilon} \rceil + 1} < \left| \frac{1}{\frac{1}{\epsilon}} \right| = |\epsilon| = \epsilon.$$

Because this is true for every $\epsilon > 0$, we have $p_n \to p$. □

This looks familiar, doesn't it? We used the same N in Example 14.4 to prove that $s_n = \frac{1}{n}$ converges to 0. What's going on? In this proof, we assembled a sequence $\{p_n\}$ of points in E, for which the nth element's distance from p is smaller than the nth element of the sequence $s_n = \frac{1}{n}$. As $s_n = \frac{1}{n}$ converges to 0, the distances between p_n and p also converge to 0.

Corollary 14.9. *(Sequences in Infinite Subsets of Compact Sets)*
Every infinite subset E of a compact set K contains a sequence that converges to a point of K.

Again, note that the limit point p may not be in E. In that case we would say $\{p_n\}$ converges in K, but not in E.

Proof. By Theorem 11.13, since K is compact, any infinite subset E of K has some limit point $p \in K$. Then by Theorem 14.8, E contains a sequence that converges to that $p \in K$. □

Corollary 14.10. *(Sequences in Infinite Bounded Real Sets)*
Every bounded infinite subset E of $\mathbb{R}^k$ contains a sequence that converges to a point of $\mathbb{R}^k$.

Proof. By the Weierstrass theorem (Theorem 12.8), since E is bounded and infinite in $\mathbb{R}^k$, it has a limit point $p \in \mathbb{R}^k$. Then by Theorem 14.8, E contains a sequence that converges to that $p \in \mathbb{R}^k$. □

As these last corollaries showed, sequences are hiding everywhere! They interact nicely with the topology we studied, and sequences of sums (called *series*) will be involved in the computation of integrals.

Although convergence might seem like a narrow topic at first, it turns out to be a rich and widely applicable idea. In the next chapter, we'll explore sequences in more detail. (In fact, that's what we'll be doing for the next four chapters. Yay, so fun!)

CHAPTER 15

Limits and Subsequences

In this chapter, we will first examine how limits of sequences interact with algebraic operations, such as addition. Then, we will study *subsequences*, which are pretty much what they sound like: the analog of subsets, but for sequences.

It turns out that when we add two convergent sequences together, the new sequence still converges, and its limit is simply the sum of the two original limits. The same is true for scalar multiplication, multiplication, and division.

Of course, we cannot apply these operations (addition, etc.) in an arbitrary metric space X, since X may not support them (remember, the only operation a metric space must have is a distance function d). So we will restrict these properties to $\mathbb{R}^k$ or $\mathbb{C}$.

Actually, for the sake of simplicity, we'll start by only proving them for $\mathbb{R}$. (It will be easy to generalize later.)

Theorem 15.1. *(Algebraic Operations on Limits in $\mathbb{R}$)*
For any sequences $\{s_n\}$ and $\{t_n\}$ in $\mathbb{R}$, if $\lim_{n\to\infty} s_n = s$ and $\lim_{n\to\infty} t_n = t$, then

1. $\lim_{n\to\infty}(s_n + t_n) = s + t$.
2. $\lim_{n\to\infty} c s_n = cs$ and $\lim_{n\to\infty}(c + s_n) = c + s$, for any $c \in \mathbb{R}$.
3. $\lim_{n\to\infty} s_n t_n = st$.
4. $\lim_{n\to\infty} \frac{1}{s_n} = \frac{1}{s}$, as long as $s \neq 0$ and $s_n \neq 0$ for any $n \in \mathbb{N}$.

Proof. For each sequence, we want to prove that it converges to the correct limit.

1. We'll use the same trick as in the proof of Theorem 14.6. Given any $\epsilon > 0$, because $\{s_n\}$ and $\{t_n\}$ converge, there exist $N_1, N_2 \in \mathbb{N}$ such that

$$n \geq N_1 \implies |s_n - s| < \frac{\epsilon}{2},$$

$$n \geq N_2 \implies |t_n - t| < \frac{\epsilon}{2}.$$

Letting $N = \max\{N_1, N_2\}$, we have that for all $n \geq N$,

$$|(s_n + t_n) - (s + t)| = |(s_n - s) + (t_n - t)|$$

$$\leq |s_n - s| + |t_n - t| < \frac{\epsilon}{2} + \frac{\epsilon}{2} = \epsilon.$$

Since this is true for every $\epsilon > 0$, we have $s_n + t_n \to s + t$.

2. If $c = 0$, we have it easy. $cs_n = 0$ for any $n \geq 0$, so given an $\epsilon > 0$, we have

$$|cs_n - cs| = |0 - 0| = 0 < \epsilon.$$

Otherwise, given any $\epsilon > 0$, because $\{s_n\}$ converges, there exists $N \in \mathbb{N}$ such that

$$n \geq N \implies |s_n - s| < \frac{\epsilon}{|c|}.$$

Then for all $n \geq N$, we have

$$|cs_n - cs| = |c|\,|s_n - s| < |c|\left(\frac{\epsilon}{|c|}\right) = \epsilon.$$

Since this is true for every $\epsilon > 0$, we have $cs_n \to cs$.

And given any $\epsilon > 0$, because $\{s_n\}$ converges, there exists $N \in \mathbb{N}$ such that

$$n \geq N \implies |(c + s_n) - (c + s)| = |(c - c) + (s_n - s)| = |s_n - s| < \epsilon.$$

Since this is true for every $\epsilon > 0$, we have $c + s_n \to c + s$. (Alternatively, we could just apply Property 1 using the constant sequence $t_n = c$.)

3. This one is a little trickier. We can make $|s_n - s|$ and $|t_n - t|$ arbitrarily small, so using $\sqrt{\epsilon}$, we could get $|s_n - s|\,|t_n - t| < (\sqrt{\epsilon})(\sqrt{\epsilon}) = \epsilon$. What we really need, though, is to show $|s_n t_n - st| < \epsilon$.

By simplifying the first product, we derive the identity

$$\begin{aligned}
|s_n - s|\,|t_n - t| &= |(s_n - s)(t_n - t)| \\
&= |s_n t_n - st_n - ts_n + st| \\
&= |(s_n t_n - st) - (st_n + ts_n - 2st)| \\
&\geq |s_n t_n - st| - |st_n + ts_n - 2st|,
\end{aligned}$$

so that

$$|s_n t_n - st| \leq |s_n - s|\,|t_n - t| + |st_n + ts_n - 2st|.$$

This is exactly what we want—but there's that pesky $|st_n + ts_n - 2st|$ term added on.

Well, by applying the previous two facts, we know that $st_n \to st$ and $ts_n \to ts$ (since s, t are constant numbers in $\mathbb{R}$), so

$$st_n + ts_n - 2st \to st + ts - 2st = 0.$$

In other words, for every $\epsilon > 0$ and for $n \geq$ some $N \in \mathbb{N}$, we have $|st_n + ts_n - 2st| < \epsilon$.

Here is the formal proof. Given any $\epsilon > 0$, because $\{s_n\}$ and $\{t_n\}$ converge, there exist $N_1, N_2 \in \mathbb{N}$ such that

$$n \geq N_1 \implies |s_n - s| < \sqrt{\frac{\epsilon}{2}},$$

$$n \geq N_2 \implies |t_n - t| < \sqrt{\frac{\epsilon}{2}}.$$

Similarly, because $st_n + ts_n - 2st \to st + ts - 2st = 0$, there exists $N_3 \in \mathbb{N}$ such that
$$n \geq N_3 \implies |st_n + ts_n - 2st| < \frac{\epsilon}{2}.$$

Letting $N = \max\{N_1, N_2, N_3\}$, we have that for all $n \geq N$,
$$|s_n t_n - st| \leq |s_n - s||t_n - t| + |st_n + ts_n - 2st|$$
$$< \left(\sqrt{\frac{\epsilon}{2}}\right)\left(\sqrt{\frac{\epsilon}{2}}\right) + \frac{\epsilon}{2} = \epsilon.$$

Since this is true for every $\epsilon > 0$, we have $s_n t_n \to st$.

4. Speaking in nonprecise terms, we can use $|s_n - s| < \epsilon$, and we need to show $\left|\frac{1}{s_n} - \frac{1}{s}\right| < \epsilon$. Note that $\left|\frac{1}{s_n} - \frac{1}{s}\right| = |s_n - s|\left|\frac{1}{s_n s}\right|$. The limit s is just a constant number that will be easy to eliminate, so we need a way to get rid of $|s_n|$. Notice that
$$|s| - |s_n| \leq |s - s_n| = |s_n - s| < \epsilon,$$
so if we choose $\epsilon = \frac{1}{2}|s|$, then $|s_n| > \frac{1}{2}|s|$.

We end up with
$$\left|\frac{1}{s_n} - \frac{1}{s}\right| = |s_n - s|\left|\frac{1}{s_n s}\right| < \epsilon \frac{2}{|s|^2},$$

so we want to choose ϵ to actually be $\frac{1}{2}|s|^2 \epsilon$.

Here is the formal proof. Because $\{s_n\}$ converges, there exists $N_1 \in \mathbb{N}$ such that
$$n \geq N_1 \implies |s_n - s| < \frac{1}{2}|s|,$$

so that $|s_n| > \frac{1}{2}|s|$. Similarly, given any $\epsilon > 0$, there exists $N_2 \in \mathbb{N}$ such that
$$n \geq N_2 \implies |s_n - s| < \frac{1}{2}|s|^2 \epsilon.$$

Letting $N = \max\{N_1, N_2\}$, we have that for all $n \geq N$,
$$\left|\frac{1}{s_n} - \frac{1}{s}\right| = |s_n - s|\left|\frac{1}{s_n s}\right|$$
$$< \left(\frac{1}{2}|s|^2 \epsilon\right)\left(\frac{2}{|s|^2}\right) = \epsilon.$$

Since this is true for every $\epsilon > 0$, we have $\frac{1}{s_n} \to \frac{1}{s}$. $\square$

To generalize these results to $\mathbb{R}^k$, we must first study the convergence of real vectors.

Theorem 15.2. *(Convergence of Real Vectors)*
Let $\mathbf{x}_n = (\alpha_{1_n}, \alpha_{2_n}, \ldots, \alpha_{k_n})$ be a vector in $\mathbb{R}^k$. Then $\{\mathbf{x}_n\}$ converges to $\mathbf{x} = (\alpha_1, \alpha_2, \ldots, \alpha_k)$ if and only if $\lim_{n \to \infty} \alpha_{j_n} = \alpha_j$ for every j between 1 and k.

In other words, saying that a sequence of real vectors converges to $\mathbf{x}$ is equivalent to saying: for every dimension $1 \leq j \leq k$, the sequence made up by taking the jth component of each vector in the sequence, converges to the jth component of $\mathbf{x}$. In symbols, this means:

$$\lim_{n \to \infty} (\alpha_{1_n}, \alpha_{2_n}, \ldots, \alpha_{k_n}) = \left(\lim_{n \to \infty} \alpha_{1_n}, \lim_{n \to \infty} \alpha_{2_n}, \ldots, \lim_{n \to \infty} \alpha_{k_n} \right).$$

This result seems obvious. If k sequences each converge, then the vectors made up of all those k sequences also converge. For example, if $\mathbf{x}_n = (\frac{1}{n}, 3)$ then $\mathbf{x}_n \to (0, 3)$. If, however $\mathbf{x}_n = (\frac{1}{n}, 3, n^2)$, then $\mathbf{x}_n$ does not converge—because even though its first two components converge, the third component n^2 does not converge.

Proof. If $\mathbf{x}_n \to \mathbf{x}$, then for every $\epsilon > 0$, there exists $N \in \mathbb{N}$ such that

$$n \geq N \implies |\mathbf{x}_n - \mathbf{x}| < \epsilon.$$

By Definition 6.10, we see that for any j between 1 and k,

$$|\alpha_{j_n} - \alpha_j| = \sqrt{|\alpha_{j_n} - \alpha_j|^2}$$

$$< \sqrt{\sum_{j=1}^{k} |\alpha_{j_n} - \alpha_j|^2}$$

$$= |\mathbf{x}_n - \mathbf{x}| < \epsilon,$$

for any $\epsilon > 0$ and for any $n \geq N$. Thus $\alpha_{j_n} \to \alpha_j$ for any j between 1 and k.

To prove the other direction, assume $\alpha_{j_n} \to \alpha_n$ for every j between 1 and k. Fill in the blanks for the rest of the proof below. The trickiest part will be deciding what to use for ϵ; one thing you could try is filling in the blanks in Box 15.1 in backward order so you can figure out what it should be.

BOX 15.1

PROVING THE OTHER DIRECTION OF THEOREM 15.2

For every j between 1 and k, we have that for every $\epsilon > 0$, there exists _____ such that

$$n \geq N_j \implies |\alpha_{j_n} - \alpha_j| < \underline{}.$$

142 • Chapter 15

> Letting $N = \max\{N_1, N_2, \ldots, N_k\}$, we have that for any _____ $\geq N$,
>
> $$|\mathbf{x}_n - \mathbf{x}| = \sqrt{\sum_{j=1}^{k} |\alpha_{j_n} - \alpha_j|^2}$$
>
> $$< \sqrt{\sum_{j=1}^{k} (\underline{})^2}$$
>
> $$= \sqrt{\underline{}} = \epsilon,$$
>
> for any $\epsilon > 0$ and for any $n \geq N$. Thus $\mathbf{x}_n \to $ _____.

$\square$

Using this theorem, we can now generalize Theorem 15.1 to prove some algebraic properties for limits in any Euclidean space.

Theorem 15.3. *(Algebraic Operations on Limits in $\mathbb{R}^k$)*
For any sequences $\{\mathbf{x}_n\}$ and $\{\mathbf{y}_n\}$ in $\mathbb{R}^k$, if $\lim_{n\to\infty} \mathbf{x}_n = \mathbf{x}$ and $\lim_{n\to\infty} \mathbf{y}_n = \mathbf{y}$, then

1. $\lim_{n\to\infty}(\mathbf{x}_n + \mathbf{y}_n) = \mathbf{x} + \mathbf{y}$.
2. $\lim_{n\to\infty} \beta_n \mathbf{x}_n = \beta \mathbf{x}$ *for any sequence $\{\beta_n\}$ in $\mathbb{R}$ that converges to β.*
3. $\lim_{n\to\infty}(\mathbf{x}_n \cdot \mathbf{y}_n) = \mathbf{x} \cdot \mathbf{y}$ *(using the scalar product, as defined in Definition 6.10).*

⚑ Note the following two differences between these properties, and those of limits in $\mathbb{R}^2$. First, scalar multiplication works not just for scalars, but also for sequences of scalars. (If you ever want plain scalar multiplication, you can just let $\beta_n = c$ for every $n \in \mathbb{N}$.) Second, there is no property for division of limits, since there is no analog of division in $\mathbb{R}^k$ when $k > 1$ (although there is division in $\mathbb{C}$).

Proof. Let's set $\mathbf{x}_n = (x_{1_n}, x_{2_n}, \ldots, x_{k_n})$ and $\mathbf{y}_n = (y_{1_n}, y_{2_n}, \ldots, y_{k_n})$, and set $\mathbf{x} = (x_1, x_2, \ldots, x_k)$ and $\mathbf{y} = (y_1, y_2, \ldots, y_k)$. By Theorem 15.2, because we have $\mathbf{x}_n \to \mathbf{x}$ and $\mathbf{y}_n \to \mathbf{y}$, we know $x_{j_n} \to x_j$ and $y_{j_n} \to y_j$ for any j between 1 and k.

1. By Property 1 of Theorem 15.1, we know $x_{j_n} + y_{j_n} \to x_j + y_j$ for any j between 1 and k. Then

$$\lim_{n\to\infty}(\mathbf{x}_n + \mathbf{y}_n) = \lim_{n\to\infty}(x_{1_n} + y_{1_n}, x_{2_n} + y_{2_n}, \ldots, x_{k_n} + y_{k_n})$$
$$= \left(\lim_{n\to\infty}(x_{1_n} + y_{1_n}), \lim_{n\to\infty}(x_{2_n} + y_{2_n}), \ldots, \lim_{n\to\infty}(x_{k_n} + y_{k_n})\right)$$
(by Theorem 15.2)
$$= (x_1 + y_1, x_2 + y_2, \ldots, x_k + y_k) = \mathbf{x} + \mathbf{y}.$$

2. By Property 3 of Theorem 15.1, we know $\beta_n x_{j_n} \to \beta x_j$ for any j between 1 and k. Then

$$\lim_{n\to\infty} \beta_n \mathbf{x}_n = \lim_{n\to\infty} (\beta_n x_{1_n}, \beta_n x_{2_n}, \ldots, \beta_n x_{k_n})$$
$$= \left(\lim_{n\to\infty} \beta_n x_{1_n}, \lim_{n\to\infty} \beta_n x_{2_n}, \ldots, \lim_{n\to\infty} \beta_n x_{n_k}\right)$$
(by Theorem 15.2)
$$= (\beta x_1, \beta x_2, \ldots, \beta x_k) = \beta \mathbf{x}.$$

3. This proof is just like the others. Fill in the blanks in Box 15.2.

BOX 15.2

> PROVING PROPERTY 3 OF THEOREM 15.3
>
> By Property 3 of Theorem 15.1, we know $x_{j_n} y_{j_n} \to$ _____ for any $1 \leq j \leq$ _____. Then
>
> $$\lim_{n\to\infty} (\mathbf{x}_n \cdot \mathbf{y}_n) = \lim_{n\to\infty} (x_{1_n} y_{1_n} + x_{2_n} y_{2_n} + \ldots + x_{k_n} y_{k_n})$$
> $$= \underline{\qquad\qquad} + \underline{\qquad\qquad} + \ldots + \underline{\qquad\qquad}$$
> (by Property 1 of _____)
> $$= x_1 y_1 + x_2 y_2 + \ldots + x_k y_k = \underline{\qquad\qquad}.$$

□

Let's change gears to something a bit less technical, and hopefully more interesting.

Definition 15.4. *(Subsequence)*
*For any sequence $\{p_n\}$ in a metric space X, let $\{n_k\}$ be a sequence of natural numbers, with $n_1 < n_2 < \ldots$. Then the sequence given by $\{p_{n_k}\}$ is a **subsequence** of $\{p_n\}$.*
*If $\{p_{n_k}\}$ converges to some $p \in X$, then p is a **subsequential limit** of $\{p_n\}$.*

 Here, the sequence n_k is just a list of increasing natural numbers. For example, if $\{n_k\} = 1, 3, 100, \ldots$, then $\{p_{n_k}\} = p_1, p_3, p_{100}, \ldots$. The list of indexes n_k is not the subsequence in question; the list of elements p_{n_k} is the subsequence.

Since $\{n_k\}$ is a sequence, it must be infinite. Therefore, something like $p_1, p_3, p_{100}, \ldots$ is a subsequence of $\{p_n\}$, whereas something like p_1, p_3, p_{100} is not (note the absence of ellipses indicates that this list does not go on to infinity, so the indexes 1, 3, 100 are not themselves a sequence). This means that just like sequences, subsequences must go on to infinity.

Example 15.5. *(Subsequences)*
Consider all of these examples in the metric space $\mathbb{R}$ or $\mathbb{C}$.

1. If $\{s_n\} = 1$ for every $n \in \mathbb{N}$, take $n_k = 2k - 1$ for every $k \in \mathbb{N}$. Then the subsequence $\{s_{n_k}\} = s_1, s_3, s_5, \ldots$ is just $1, 1, 1, \ldots$, which is the same as $\{s_n\}$ itself.

2. If $\{s_n\} = i^n$ for every $n \in \mathbb{N}$, take $n_k = 4k$ for every $k \in \mathbb{N}$. Then the subsequence $\{s_{n_k}\} = s_4, s_8, s_{12}, \ldots$ is just $1, 1, 1, \ldots$, which converges to the point 1. So 1 is a subsequential limit of $\{s_n\}$, even though it is not a limit of $\{s_n\}$.
3. If $\{p_n\} = (-1)^n + \frac{(-1)^n}{n}$ for every $n \in \mathbb{N}$, take $n_k = 2k$ for every $k \in \mathbb{N}$. Then the subsequence $\{p_{n_k}\} = p_2, p_4, p_6, \ldots$ is $\frac{3}{2}, \frac{5}{4}, \frac{7}{6}, \ldots$, meaning $\{p_{n_k}\} = 1 + \frac{1}{2k}$. This subsequence converges to the point 1. So 1 is a subsequential limit of $\{p_n\}$, even though it is not a limit of $\{p_n\}$.

Similarly, by taking $n_k = 2k - 1$, we obtain the subsequence $\{p_{n_k}\} = p_1, p_3, p_5, \ldots$ This subsequence converges to -1, so -1 is also a subsequential limit of $\{p_n\}$.

Thus, if $\{p_n\}$ "alternates" between two almost-limits, it does not converge, but these almost-limits *are* subsequential limits. (Take another look at Figures 14.4 and 14.5.) This concept will be explored in Chapter 17 in much greater detail.

Theorem 15.6. *(Converges $\iff$ Every Subsequence Converges)*
For any sequence $\{p_n\}$ in a metric space X, $\{p_n\}$ converges to $p \in X$ if and only if every subsequence of $\{p_n\}$ converges to p.

Proof. Because a subsequence is just a subset of $\{p_n\}$'s elements, any element of the subsequence past N will still be a distance of less than ϵ away from p. To put this more formally: if $p_n \to p$, then for every $\epsilon > 0$, there exists $N \in \mathbb{N}$ such that

$$n \geq N \implies d(p_n, p) < \epsilon.$$

Because the n_k are increasing with k, $k \geq N \implies k_n \geq N$. Thus for every subsequence $\{p_{n_k}\}$ of $\{p_n\}$, we have

$$k \geq N \implies n_k \geq N \implies d(p_{n_k}, p) < \epsilon,$$

for that same ϵ and N. Since this is true for every $\epsilon > 0$, we have $p_{n_k} \to p$.

The other direction of the proof is easy. Since $\{p_n\}$ itself is a subsequence of $\{p_n\}$ (with $n_k = k$), then if every subsequence of $\{p_n\}$ converges so must $\{p_n\}$. □

This next theorem should look similar to Corollary 14.9.

Theorem 15.7. *(Subsequences in Compact Sets)*
For any sequence $\{p_n\}$ in a compact metric space X, some subsequence of $\{p_n\}$ converges to some point $p \in X$.

What is the difference between Corollary 14.9 and Theorem 15.7? The former says that any infinite set E in a compact set has a convergent sequence, but the elements of E are in no particular order (E might even be uncountable). This new theorem says that any *sequence* in a compact set has a convergent subsequence.

Letting E be the range of $\{p_n\}$, we might think that Corollary 14.9 can help us prove the case in which E is infinite. However, there is something wrong with the following argument.

If E (the range of $\{p_n\}$) is an infinite subset of the compact metric space X, then by Corollary 14.9, we know there is some sequence $\{s_n\}$ in E that converges to a point of $p \in X$. The range of $\{s_n\}$ is a subset of E, so $\{s_n\}$ is a subsequence of $\{p_n\}$. We have thus found a subsequence of $\{p_n\}$ that converges to a point of X.

Can you guess where the error is? We cannot be sure that $\{s_n\}$ is a subsequence of $\{p_n\}$, just because it is contained in $\{p_n\}$'s range E. Why? Because the points might be in a different order!

As a basic example, let $\{p_n\} = n^2$ for every $n \in \mathbb{N}$. Then $E = \{1, 4, 9, \ldots\}$, so E contains the sequence $s_n = 1, 1, 1, \ldots$. However, this $\{s_n\}$ is not a subsequence of $\{p_n\}$, since the element 1 does not occur an infinite number of times (for distinct n_k) in $\{p_n\}$.

Don't worry, the proof of this theorem doesn't use any crazy techniques we haven't seen before. Just make sure you understand why we cannot use the above method.

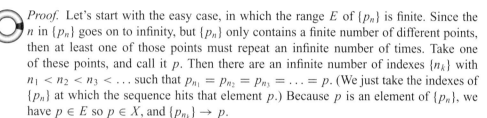

Proof. Let's start with the easy case, in which the range E of $\{p_n\}$ is finite. Since the n in $\{p_n\}$ goes on to infinity, but $\{p_n\}$ only contains a finite number of different points, then at least one of those points must repeat an infinite number of times. Take one of these points, and call it p. Then there are an infinite number of indexes $\{n_k\}$ with $n_1 < n_2 < n_3 < \ldots$ such that $p_{n_1} = p_{n_2} = p_{n_3} = \ldots = p$. (We just take the indexes of $\{p_n\}$ at which the sequence hits that element p.) Because p is an element of $\{p_n\}$, we have $p \in E$ so $p \in X$, and $\{p_{n_k}\} \to p$.

Now if E is infinite, we can apply Theorem 11.13 to obtain a limit point p of E. We will construct a subsequence $\{p_{n_i}\}$ of $\{p_n\}$ that converges to that $p \in X$.

Recall from our discussion preceding Definition 13.2 that any infinite construction should work something like a proof by induction. It must have a first step and a step specifying $i + 1$ given i.

1. Every neighborhood of p contains an infinite number of points of E. Let's start by taking $N_1(p)$, which contains some element of E. We label this element p_{n_1}, and we have $d(p_{n_1}, p) < 1$.
2. Say we already have points $p_{n_1}, p_{n_2}, \ldots, p_{n_i}$ with $n_1 < n_2 < \ldots < n_i$, such that $d(p_{n_k}, p) < \frac{1}{k}$ for any $k \leq i$. How do we find $p_{n_{i+1}}$? Well, $N_{\frac{1}{i+1}}(p)$ contains an infinite number of points of E. So even if it already contains all the points $p_{n_1}, p_{n_2}, \ldots, p_{n_i}$, it must contain at least one more. Call that new point $p_{n_{i+1}}$, and we have $d(p_{n_{i+1}}, p) < \frac{1}{i+1}$.

    How do we guarantee that $n_i < n_{i+1}$? In other words, how can we make sure that $p_{n_{i+1}}$ comes after p_{n_i} in the sequence $\{p_n\}$? Well, there are only a finite number of indexes that go up to n_i, but $N_{\frac{1}{i+1}}(p)$ contains an infinite number of points of E. So there must be at least one other index n_{i+1} that comes after n_i such that $p_{n_{i+1}} \in N_{\frac{1}{i+1}}(p)$. Thus $n_i < n_{i+1}$, and the construction can proceed for $i + 2, i + 3, i + 4, \ldots$.

Now we have a subsequence $\{p_{n_i}\}$ with $d(p_{n_i}, p) < \frac{1}{i}$ for any $i \in \mathbb{N}$. For any $\epsilon > 0$, we let $N = \lceil \frac{1}{\epsilon} \rceil + 1$, so by the same argument as in the proof of Theorem 14.8, we see $p_{n_i} \to p$. $\square$

Not to be confused with the Weierstrass theorem (Theorem 12.8), this next one is known as the Bolzano-Weierstrass theorem. (Weierstrass must have been one busy guy!)

Theorem 15.8. *(Bolzano-Weierstrass Theorem)*
For any sequence $\{p_n\}$ in $\mathbb{R}^k$, if $\{p_n\}$ is bounded then some subsequence of $\{p_n\}$ converges to some point $p \in \mathbb{R}^k$.

Proof. Let E be the range of $\{p_n\}$. Since E is bounded, then Theorem 12.5 tells us E is a subset of some k-cell I, which is compact by Theorem 12.4. Then the sequence $\{p_n\}$ is contained in the compact set I, so by Theorem 15.7, it contains a subsequence that converges to a point $p \in I$. □

Theorem 15.9. *(Set of Subsequential Limits Is Closed)*
For any sequence $\{p_n\}$ in a metric space X, the set E^* of all subsequential limits of $\{p_n\}$ is closed relative to X.

Note that E^* contains the limit of *every* convergent subsequence of $\{p_n\}$.

Proof. Let q be a limit point of E^*. We want to show that $q \in E^*$, meaning there is some subsequence $\{p_{n_i}\}$ of $\{p_n\}$ that converges to q. We'll construct $\{p_{n_i}\}$ almost identically to the subsequence in the proof of Theorem 15.7.

Before starting with the construction, we want to be able to exclude the case where $p_n = q$ for every $n \geq$ some k. If $\{p_n\}$ were such a sequence, it would look like $p_1, p_2, \ldots, p_{k-1}, q, q, q, \ldots$. Then every subsequence would converge to q, so $E^* = \{q\}$, which is closed by Example 9.26.

1. Choose $n_1 \in \mathbb{N}$ so that p_{n_1} is an element of $\{p_n\}$ with $p_{n_1} \neq q$. For later convenience, set $\delta = d(p_{n_1}, q)$.
2. Say we already have points $p_{n_1}, p_{n_2}, \ldots, p_{n_i}$ with $n_1 < n_2 < \ldots < n_i$, such that $p_{n_k} \neq q$ and $d(p_{n_k}, q) < \frac{\delta}{k}$ for any $k \leq i$.

   How do we find $p_{n_{i+1}}$? The basic idea is that for any point $x \in E^*$ that is close to q, there is a subsequence of $\{p_n\}$ that converges to x. And because q is a limit point of E^*, we can make this x arbitrarily close to q.

   To put this in rigorous terms, because q is a limit point of E^*, every one of its neighborhoods contains at least one non-q point of E^*. Then there is an $x \in E^*$ such that $d(x, q) < \frac{\delta}{2(i+1)}$. Since x is a limit of some subsequence of $\{p_n\}$, past some $N \in \mathbb{N}$, each element of that subsequence is closer than a distance of $\frac{\delta}{2(i+1)}$ to x. So if we choose $n_{i+1} > \max\{N, n_i\}$, then we have $n_i < n_{i+1}$, and $d(p_{n_{i+1}}, x) < \frac{\delta}{2(i+1)}$. Thus

$$d(p_{n_{i+1}}, q) \leq d(p_{n_{i+1}}, x) + d(x, q) < \frac{\delta}{2(i+1)} + \frac{\delta}{2(i+1)} = \frac{\delta}{i+1}.$$

Now we have a subsequence $\{p_{n_i}\}$ with $d(p_{n_i}, q) < \frac{\delta}{i}$ for any $i \geq 2$. For any $\epsilon > 0$, we let $N = \lceil \frac{\delta}{\epsilon} \rceil + 1$, so by the same argument as in the proof of Theorem 14.8, we see $p_{n_i} \to p$.

Wait, what is that δ doing everywhere? Why couldn't we have just constructed a subsequence with $d(p_{n_i}, q) < \frac{1}{i}$, as in the proof of Theorem 15.7? Well, in the first step of our construction, we could not guarantee that $d(p_{n_1}, q) < 1$, only that $d(p_{n_1}, q) = \delta > 0$. Thus we had to carry that δ throughout the construction. But the last step still works, since δ is just some constant number—so whenever $n_i \geq N = \lceil \frac{\delta}{\epsilon} \rceil + 1$, we have

$$d(p_{n_i}, q) < \frac{\delta}{\lceil \frac{\delta}{\epsilon} \rceil + 1} < \epsilon.$$

□

Hopefully these subsequences aren't tripping you up too much; if they are, reread the foregoing proofs, then try to do them on your own from memory. After delving into Cauchy sequences in the next chapter, we'll come back to explore the limits of subsequences in greater depth.

CHAPTER 16

Cauchy and Monotonic Sequences

In discussing convergence, we saw that some sequences might converge in one metric space but diverge in another. Thus, like closedness and openness, the property of convergence depends on which metric space the sequence is embedded in. So is there a property that is similar to convergence, but does not depend on the embedding metric spaces? Why yes, there is! Like compactness, being *Cauchy* is a property that holds in any metric space: a Cauchy sequence somewhere is Cauchy everywhere.

Definition 16.1. *(Cauchy Sequence)*
*For any sequence $\{p_n\}$ in a metric space X, $\{p_n\}$ is a **Cauchy** sequence if for every $\epsilon > 0$, $d(p_n, p_m) < \epsilon$ for every n and m greater than or equal to some natural number N.*
In symbols, $\{p_n\}$ is Cauchy if:

$$\forall \epsilon > 0, \exists N \in \mathbb{N} \text{ such that } n, m \geq N \implies d(p_n, p_m) < \epsilon.$$

Isn't this the same as convergence in Definition 14.3? No! The key difference is that with Cauchy sequences, $d(p_n, p_m) < \epsilon$, unlike $d(p_n, p) < \epsilon$ in convergent sequences. Cauchy sequences do not get closer and closer to one point in particular; rather, the elements get closer and closer to each other. In a Cauchy sequence, given any distance $\epsilon > 0$, there is some step past which any two elements of the sequence are a distance of less than ϵ apart.

The N_ϵ challenge works just as well here, with a slight modification. Given any $\epsilon > 0$, can you find an N such that $p_N, p_{N+1}, p_{N+2}, \ldots$ are all a distance of less than ϵ away from each other?

Example 16.2. *(Cauchy Sequences)*
Defining Cauchy sequences begs the following question: if the elements get closer and closer together, how would they ever *not* converge to a single point?

For a simple example, let $p_n = \frac{1}{n}$ for any $n \in \mathbb{N}$, and take the sequence $\{p_n\}$ in the metric space $X = \mathbb{R} \setminus \{0\}$. Then $\{p_n\}$ does not converge in X, because 0 is not an element of X. But $\{p_n\}$ is Cauchy. Why? If $n, m \geq N$, we have

$$d(p_n, p_m) = \left|\tfrac{1}{n} - \tfrac{1}{m}\right| \leq \max\left\{\tfrac{1}{n}, \tfrac{1}{m}\right\} \leq \tfrac{1}{N}.$$

For any $\epsilon > 0$, we want $1 < N\epsilon$, so just let $N = \lceil \tfrac{1}{\epsilon} \rceil + 1$, and we're good.

As a further example, let s_n be $\sqrt{2}$ to the nth decimal place, in the metric space $\mathbb{Q}$. So $s_n = 1.4, 1.41, 1.414, \ldots$ Each element of the sequence is rational, since we can write the decimals as $\tfrac{14}{10}, \tfrac{141}{100}, \tfrac{1414}{1000}, \ldots$ This sequence converges to $\sqrt{2}$, so $\{s_n\}$ does not converge in $\mathbb{Q}$. But $\{s_n\}$ is Cauchy. Why? If $n, m \geq N$, we have

$$d(p_n, p_m) \leq 10^{-N}.$$

For any $\epsilon > 0$, we want $-N < \log_{10}(\epsilon)$, so just let $N = \lceil -\log_{10}(\epsilon) \rceil + 1$, and we're good.

This might lead you to think that Cauchy sequences can only converge in metric spaces similar to $\mathbb{R}$, the key ingredient being the least upper bound property (so there are no "holes" to which Cauchy sequences might converge). And you would be right! We will soon see that $\mathbb{R}$ has a property called *completeness*, which means "every Cauchy sequence converges."

Although *Cauchy* $\implies$ *convergent* only when the metric space is complete, it turns out that *convergent* $\implies$ *Cauchy* in every case.

Theorem 16.3. *(Convergent $\implies$ Cauchy)*
For any sequence $\{p_n\}$ in a metric space X, if $\{p_n\}$ converges to some $p \in X$, then $\{p_n\}$ is a Cauchy sequence.

Proof. For every $\epsilon > 0$, there is an $N \in \mathbb{N}$ such that $d(p_n, p) < \tfrac{\epsilon}{2}$ whenever $n \geq N$. Of course, for any $m \geq N$, we also have $d(p_m, p) < \tfrac{\epsilon}{2}$. Then for any $n, m \geq N$, we can use the triangle inequality to get

$$d(p_n, p_m) \leq d(p_n, p) + d(p, p_m) < \tfrac{\epsilon}{2} + \tfrac{\epsilon}{2} = \epsilon.$$

Since this is true for every $\epsilon > 0$, $\{p_n\}$ must be a Cauchy sequence. $\square$

Before continuing with Cauchy sequences, we will first define the concept of *diameter*. Although it may seem tricky at first, it will give us a more intuitive way to look at Cauchy sequences and will help us prove some theorems that would otherwise be more difficult.

Definition 16.4. *(Diameter)*
*For any nonempty subset E of a metric space X, the **diameter** of E is the supremum of the set of distances between any possible pair of elements of E.*
    In symbols:

$$\operatorname{diam} E = \sup\{d(p, q) \mid p, q \in E\}.$$

Example 16.5. (Diameter)
Consider the following as subsets of the metric space $\mathbb{R}$:

1. If $E = \{1, 2, 3\}$, then

$$\operatorname{diam} E = \sup\{d(1,2), d(2,3), d(1,3)\} = \sup\{1, 1, 2\} = 2.$$

2. If $E = [-3, 3]$ then diam $E = 6$. This shows that diameter is basically what it sounds like: the length from one end of the set to the other end.
3. If $E = [-3, 3)$, then diam E again $= 6$, since

$$\{d(-3, p) \mid p \in [-3, 3)\} = [0, 6),$$

which has 6 as its least upper bound.

4. If $E = [0, \sqrt{2})$, then diam $E = \sqrt{2}$. Even if we consider E as a subset of the metric space $\mathbb{Q}$, the diameter of E is still $\sqrt{2}$, since the diameter need not be an element of the embedding metric space. Why? Because the diameter is the supremum of a set of distances. Recall from Definition 9.1 that any distance is a real number, so every diameter is also a real number.
5. If $E = [0, \infty)$, then E has no diameter, because the set of all distances between points of E is unbounded.
6. We can also consider a sequence of diameters. For example, let $A_n = [0, \frac{1}{n})$ for every $n \in \mathbb{N}$. Now take the sequence

$$\operatorname{diam} A_1, \operatorname{diam} A_2, \operatorname{diam} A_3, \ldots,$$

which is just $1, \frac{1}{2}, \frac{1}{3}, \ldots$. Then

$$\lim_{n \to \infty} \operatorname{diam} A_n = \lim_{n \to \infty} \frac{1}{n} = 0.$$

Note that any sequence of diameters is a sequence of positive real numbers.

The following discussion will help elucidate the upcoming theorem.

If we have any sequence $\{p_n\}$, let E_N be the range of that sequence starting from the Nth step, so $E_N = \{p_N, p_{N+1}, p_{N+2}, \ldots\}$. We can build the sequence diam E_1, diam E_2, diam $E_3, \ldots$, and see whether it converges by taking $\lim_{N \to \infty}$ diam E_N. To be clear, this is the limit of the sequence

$$\operatorname{diam}\{p_1, p_2, p_3, p_4, \ldots\}, \operatorname{diam}\{p_2, p_3, p_4, \ldots\}, \operatorname{diam}\{p_3, p_4, \ldots\}, \ldots,$$

as shown in Figure 16.1.

If $p_n \to p$, then $\lim_{N \to \infty}$ diam $E_N = 0$. Why? For every $\epsilon > 0$, there is an $N \in \mathbb{N}$ such that $d(p_n, p) < \frac{\epsilon}{2}$ whenever $n \geq N$. In other words, for every $\epsilon > 0$ there is some E_N with diam $E_N < \epsilon$. (Why do we have ϵ here, not the $\frac{\epsilon}{2}$ that we originally used? Well, it could be that $p_n = p - \frac{\epsilon}{2}$, and $p_{n+1} = p + \frac{\epsilon}{2}$. Every point is a distance of at most $\frac{\epsilon}{2}$ from p, so every pair of points is a distance of at most ϵ away from each other.) Then for any $\epsilon > 0$, there is a point diam E_N of the sequence $\{\operatorname{diam} E_n\}$ that is less than ϵ.

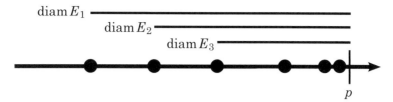

Figure 16.1. Given the sequence $\{p_n\}$ (whose first few elements are represented by the dots on the line) which converges to p, the first few elements of the sequence $\{\text{diam } E_n\}$ are shown.

In fact, this is true for all E_n with $n \geq N$, because we know that $E_n \subset E_N$, so

$$\{d(p,q) \mid p,q \in E_n\} \subset \{d(p,q) \mid p,q \in E_N\}$$

$$\implies \sup\{d(p,q) \mid p,q \in E_n\} \leq \sup\{d(p,q) \mid p,q \in E_N\}$$

$$\implies \text{diam } E_n \leq \text{diam } E_N,$$

so diam E_n is also $< \epsilon$. Therefore, $\lim_{N \to \infty} \text{diam } E_N = 0$.

Again, we can only be sure this is true if $\{p_n\}$ converges. But wait—surprise! The following theorem asserts this for any Cauchy sequence as well.

Theorem 16.6. *(Diameter of Cauchy Sequences)*
For any sequence $\{p_n\}$ in a metric space X, let E_N be the range of the subsequence $p_N, p_{N+1}, p_{N+2}, \ldots$. Then $\{p_n\}$ is a Cauchy sequence if and only if $\lim_{N \to \infty} \text{diam } E_N = 0$.

This theorem should make intuitive sense: both statements mean that the distances between points of $\{p_n\}$ become arbitrarily small.

Proof. Let's start by assuming $\{p_n\}$ is Cauchy. This proof should be similar to the argument we made in the preceding discussion. We know that for every $\epsilon > 0$, there is an $N \in \mathbb{N}$ such that $d(p_n, p_m) < \epsilon$ whenever $n, m \geq N$. Then for every $\epsilon > 0$, we have an N such that

$$\text{diam } E_N = \sup\{d(p_n, p_m) \mid p_n, p_m \in E_N\}$$

$$= \sup\{d(p_n, p_m) \mid n, m \geq N\}.$$

We know that ϵ is an upper bound of that set, so diam $E_N \leq \epsilon$. Because E_{N+1} is just the set E_N without the point p_N, we have $E_{N+1} \subset E_N$ so diam $E_{N+1} \leq E_N$ (remember $A \subset B \implies \sup A \leq \sup B$). Thus diam $E_n \leq \epsilon$ for every $n \geq N$. Therefore, for every $\epsilon > 0$, there is an $N \in \mathbb{N}$ such that $|\text{diam } E_n - 0| \leq \epsilon$ whenever $n \geq N$, so $\lim_{N \to \infty} \text{diam } E_N = 0$.

(Wait, is it a problem that we have $\leq \epsilon$, not $< \epsilon$? Not really, since for the first step, we could have just taken an N such that $d(p_n, p_m) < \frac{\epsilon}{2}$, so that diam $E_N \leq \frac{\epsilon}{2} < \epsilon$, and we would have $|\text{diam } E_n - 0| < \epsilon$. If we were writing a 100 percent precise proof, we would have done that, but it would have made it slightly more confusing.)

Conversely, assume diam $E_N \to 0$. Then for every $\epsilon > 0$, there is an $N \in \mathbb{N}$ such that $|\text{diam } E_n - 0| < \epsilon$ whenever $n \geq N$. So

$$\epsilon \geq \sup \{d(p_n, p_m) \mid n, m \geq N\},$$

meaning $d(p_n, p_m) \leq \epsilon$ whenever $n, m \geq N$. Thus $\{p_n\}$ is Cauchy. □

The following two properties of diameter will help us prove crucial theorems for Cauchy sequences.

Theorem 16.7. *(Diameter of a Closure)*
For any subset E of a metric space X, diam $\overline{E}$ = diam E.

Proof. To assert equality, we will prove diam $\overline{E} \geq$ diam E and then diam $\overline{E} \leq$ diam E. The first one is easy, since $E \subset \overline{E} \implies$ diam $E \leq$ diam $\overline{E}$.

For the other inequality, if we can show diam $\overline{E} \leq$ diam $E + \epsilon$ for every choice of $\epsilon > 0$, then we will have diam $\overline{E} \leq$ diam E. (Since if diam $\overline{E}$ were $>$ diam E, there would be some $c \neq 0$ with diam $\overline{E}$ = diam $E + c$, so for $\epsilon = \frac{c}{2}$, we would have diam $\overline{E} >$ diam $E + \epsilon$, which is a contradiction.)

For any $\epsilon > 0$, take any point p of $\overline{E}$. Either $p \in E$, or else p is a limit point of E. If $p \in E$, let $p' = p$, so $d(p, p') = 0 < \frac{\epsilon}{2}$. If p is a limit point of E, then there is a point $p' \neq p$ of E inside $N_{\frac{\epsilon}{2}}(p)$, so that $d(p, p') < \frac{\epsilon}{2}$. Similarly, for any $q \in \overline{E}$, we can find a $q' \in E$ with $d(q, q') < \frac{\epsilon}{2}$. Then

$$d(p, q) \leq d(p, p') + d(p', q) \quad \text{(triangle inequality)}$$
$$\leq d(p, p') + d(p', q') + d(q', q) \quad \text{(triangle inequality again)}$$
$$< \frac{\epsilon}{2} + d(p', q') + \frac{\epsilon}{2}$$
$$\leq \text{diam } E + \epsilon.$$

The last step is true since $p', q' \in E$, and diam E is an upper bound of the set of all distances in E. Since $d(p, q) \leq$ diam $E + \epsilon$ for every possible pair of points p and q in $\overline{E}$, we have diam $\overline{E} \leq$ diam $E + \epsilon$. □

Theorem 16.8. *(Diameter of Nested Compact Sets)*
Let $\{K_n\}$ be a collection of nonempty compact sets in a metric space X such that for any $n \in \mathbb{N}$, $K_n \supset K_{n+1}$. If $\lim_{n \to \infty}$ diam $K_n = 0$, then $\bigcap_{n=1}^{\infty} K_n$ contains exactly one point.

Proof. Fill in the blanks for this simple proof in Box 16.1.

Cauchy and Monotonic Sequences • 153

BOX 16.1

PROVING THEOREM 16.8

By _____, we know that $\bigcap_{n=1}^{\infty} K_n$ contains at least one point. If it contains more then one point, say, p and q, let $r = d(p,q)$, so diam $\bigcap_{n=1}^{\infty} K_n \geq r > 0$. Then for any $m \in \mathbb{N}$, $\bigcap_{n=1}^{\infty} K_n \subset$ _____ implies

$$\text{diam } K_m \underline{\qquad} \text{diam } \bigcap_{n=1}^{\infty} K_n = r.$$

But _____ $\to 0$, which is impossible if each element of that sequence is $\geq r$, since r is a fixed positive number.

We are now equipped to prove the converse of Theorem 16.3 for two special cases: all Cauchy sequences converge in compact metric spaces, and all Cauchy sequences converge in $\mathbb{R}^k$.

Theorem 16.9. *(Cauchy $\Longrightarrow$ Convergent in Compact Sets)*
If $\{p_n\}$ is a Cauchy sequence in a compact metric space X, then $\{p_n\}$ converges to some $p \in X$.

Proof. Let E_N be the range of the subsequence $p_N, p_{N+1}, p_{N+2}, \ldots$, so by Theorem 16.6, $\lim_{N \to \infty} \text{diam } E_N = 0$. To make use of Theorem 16.8, we want some sequence of nested compact sets involving E_N. Since each $\overline{E_N}$ is a subset of the compact set X, each $\overline{E_N}$ is itself compact by Theorem 11.8. For every $N \in \mathbb{N}$, $E_N \supset E_{N+1}$ implies $\overline{E_N} \supset \overline{E_{N+1}}$ (by Corollary 10.9), and by Theorem 16.7, we also have

$$\lim_{N \to \infty} \text{diam } \overline{E_N} = \lim_{N \to \infty} \text{diam } E_N = 0.$$

Then Theorem 16.8 tells us that there is exactly one point p in the intersection $\bigcap_{N=1}^{\infty} \overline{E_N}$. We will show that $p_n \to p$.

For any $\epsilon > 0$, there is an $N \in \mathbb{N}$ such that $d(\text{diam } \overline{E_n}, 0) < \epsilon$ whenever $n \geq N$, so diam $\overline{E_N} < \epsilon$ whenever $n \geq N$ (since any diameter is just a nonnegative real number). Because $p \in \overline{E_n}$, we know that for any $q \in \overline{E_n}$

$$d(p,q) \leq \sup \{d(p,q) \mid p, q \in \overline{E_N}\}$$
$$= \text{diam } \overline{E_N}$$
$$< \epsilon.$$

As long as $n \geq N$, the above is true for every $q \in \overline{E_n}$, so it is true for every $q \in E_n$, so it is true for every p_n. Thus $d(p, p_n) < \epsilon$ for any $n \geq N$. Since this is true for every $\epsilon > 0$, we have $p_n \to p$. □

Wait, why did we need to use Theorem 16.8 at all? Couldn't we have just used Corollary 11.12 to show that $\bigcap_{N=1}^{\infty} \overline{E_N}$ contains *at least* (rather than *exactly*) one point p, and then $p_n \to p$? Yes, we could have. (In fact, we could have then proven that

$\bigcap_{N=1}^{\infty} \overline{E_N}$ can only contain one element in this case. Take a point $p' \in \bigcap_{N=1}^{\infty} \overline{E_N}$, and use the same argument as for p to show that $p_n \to p'$, so by Theorem 14.6 we have $p' = p$.) But proving Theorem 16.8 was still a useful exercise—and I'm sure you're just as glad as I am that we did it, right?

Theorem 16.10. *(Cauchy $\implies$ Convergent in $\mathbb{R}^k$)*
For any sequence $\{\mathbf{x}_n\}$ in the metric space $\mathbb{R}^k$, if $\{\mathbf{x}_n\}$ is a Cauchy sequence, then $\{\mathbf{x}_n\}$ converges to some $\mathbf{x} \in \mathbb{R}^k$.

Proof. If we can show that $\{\mathbf{x}_n\}$ is bounded, then by Theorem 12.5, the range of $\{\mathbf{x}_n\}$ would be contained in some k-cell I, which is compact by Theorem 12.4. Thus $\{\mathbf{x}_n\}$ would be contained in a compact set, so it would converge by Theorem 16.9.

Let E_N be the range of the subsequence $\mathbf{x}_N, \mathbf{x}_{N+1}, \mathbf{x}_{N+2}, \ldots$, so by Theorem 16.6, $\lim_{N \to \infty}$ diam $E_N = 0$. Then there is an $N \in \mathbb{N}$ with $d(\text{diam } E_N, 0) < 1$, so the distance between any two points of E_N is less than 1, so E_N is bounded. Notice that the range of $\{\mathbf{x}_n\}$ is the set $\{\mathbf{x}_1, \mathbf{x}_2, \ldots, \mathbf{x}_{N-1}, \mathbf{x}_N, \mathbf{x}_{N+1}, \ldots\}$, which is just $\{\mathbf{x}_1, \mathbf{x}_2, \ldots, \mathbf{x}_{N-1}\} \cup E_N$. Because $\{\mathbf{x}_1, \mathbf{x}_2, \ldots, \mathbf{x}_{N-1}\}$ is finite, we can bound it by the maximum distance between two of its points. Thus the range of $\{\mathbf{x}_n\}$ is the union of two bounded sets, so $\{\mathbf{x}_n\}$ is bounded by Theorem 9.5, so $\{\mathbf{x}_n\}$ indeed converges. $\square$

Definition 16.11. *(Complete)*
*A metric space X is **complete** if every Cauchy sequence in X converges to some point of X.*

Example 16.12. (Complete Metric Spaces)
By Theorem 16.9, any compact metric space is complete. By Theorem 16.10, any Euclidean space $\mathbb{R}^k$ is a complete metric space.

On the other hand, the metric space $\mathbb{Q}$ is *not* complete, since Example 16.2 provides a rational Cauchy sequence that does not converge to any point of $\mathbb{Q}$.

As we noticed in Example 16.2, it looks like the least upper bound property is what helps all Cauchy sequences in $\mathbb{R}$ converge. In fact, it turns out that for any ordered field F, the following two statements are equivalent:

Statement 1. F has the least upper bound property.
Statement 2. F is complete and F has the Archimedean property.

Note that F must be an ordered field—not only a metric space—so that we can have a meaningful definition of upper bounds. (Remember that all ordered fields are metric spaces, with the distance function $d(p,q) = |p - q|$.)

The proof of this equivalence is pretty long and boring, so I'll let you off the hook. You can use the time it would have taken you to understand the proof to bake a pie instead! (Note that pies are complete, since any Cauchy sequence of filling always converges to a crusty limit.)

Theorem 16.13. *(Closed Subsets of Complete Metric Spaces)*
Let E be a subset of a complete metric space X. If E is closed, then E is also complete.

Proof. Let $\{p_n\}$ be a Cauchy sequence in E. Then $\{p_n\}$ is also in X, so it converges to some point $p \in X$. By the alternate definition of convergence in Theorem 14.5, we know that every neighborhood of p contains infinitely many points of E, so p is a limit point of E. (This excludes the constant-tail case where $p_n = p$ for all n large enough, in which case $p_n \to p$ and we are done.) Because E is closed, we have $p \in E$, so $\{p_n\}$ converges in E. Thus E is complete. □

We will now change gears to learn about a different type of sequence, called *monotonic*.

Definition 16.14. *(Monotonic Sequence)*
For any sequence $\{s_n\}$ in an ordered field F, $\{s_n\}$ is **monotonically increasing** if $s_n \leq s_{n+1}$ for every $n \in \mathbb{N}$; and $\{s_n\}$ is **monotonically decreasing** if $s_n \geq s_{n+1}$ for every $n \in \mathbb{N}$.

If $\{s_n\}$ is monotonically increasing and/or decreasing, we say $\{s_n\}$ is a **monotonic** (or **monotone**) sequence.

Example 16.15. (Monotonic Sequences)
The sequence $1, 2, 3, \ldots$ is monotonically increasing. The sequence given by $s_n = \frac{1}{n}$ for every $n \in \mathbb{N}$ is monotonically decreasing. The sequence $1, 1, 1, \ldots$ is both monotonically increasing and decreasing.

In Theorem 14.7, we saw that all convergent sequences are bounded. We know that the converse, however, is not necessarily true ($s_n = (-1)^n$ is bounded but not convergent).

It turns out that the converse does hold for monotonic sequences in ordered fields that have the least upper bound property. Thus in those fields, both Cauchy sequences and bounded monotonic sequences are guaranteed to converge.

Theorem 16.16. *(Bounded Monotonic Sequences)*
Let $\{s_n\}$ be a monotonic sequence in an ordered field F, and let F have the least upper bound property. Then $\{s_n\}$ converges in F if and only if $\{s_n\}$ is bounded.

Proof. We already have one direction of the proof: if $\{s_n\}$ converges, then it is bounded by Theorem 14.7.

Otherwise, the main idea is to take the least upper bound s of a bounded sequence $\{s_n\}$, and show that there will always be elements of $\{s_n\}$ between s and $s - \epsilon$ (otherwise s is not the supremum).

For the formal proof, take a bounded sequence $\{s_n\}$, and assume for this case that $\{s_n\}$ is monotonically increasing. Let E be the range of $\{s_n\}$, so E is bounded. Then by Theorem 9.6, E is bounded above, so because F has the least upper bound property, $s = \sup E$ exists in F.

Pick an $\epsilon > 0$. Because s is the *least* upper bound, there must be an element of E between $s - \epsilon$ and s (otherwise $s - \epsilon$ would be an upper bound of E). So there is some $N \in \mathbb{N}$ with $s - \epsilon < s_N \leq s$.

Now for any $n \geq N$, we have $s_n \geq s_N$ (since $\{s_n\}$ is monotonically increasing), but s_n is still $\leq s$, since s is an upper bound of E. Then for all $n \geq N$,

$$s - \epsilon < s_N \leq s_n \leq s < s + \epsilon,$$

meaning $d(s_n, s) < \epsilon$.

The proof for the case that $\{s_n\}$ is monotonically *decreasing* is basically the same. Fill in the blanks in Box 16.2.

156 • Chapter 16

BOX 16.2

PROVING THEOREM 16.16 FOR MONOTONICALLY DECREASING SEQUENCES

Take any bounded, monotonically decreasing sequence $\{s_n\}$ in F. Let E be the range of $\{s_n\}$, so E is bounded below. Because F has the least upper bound property, Theorem 4.13 tells us that F also has the _____ property, so $s = $ _____ exists in F.

Pick an $\epsilon > 0$. Because s is the *greatest* lower bound, there must be an element of E between s and _____ (otherwise _____ would be a lower bound of _____). So there is some $N \in \mathbb{N}$ with _____ $\leq s_N < s + \epsilon$.

Now for any $n \geq N$, we have s_n _____ s_N (since $\{s_n\}$ is monotonically decreasing), but s_n is still $\geq s$, since s is a _____ of E. Then for all $n \geq N$,

$$s - \epsilon < s \leq s_n \leq s_N < s + \epsilon,$$

meaning $d(s_n, s) < $ _____.

☐

That was everything you ever wanted to know about Cauchy and monotonic sequences and more. Remember the following main takeaways from this chapter: any compact metric space and any Euclidean space is complete, meaning every Cauchy sequence converges, and in any ordered field with the least upper bound property (such as $\mathbb{R}$), every bounded monotonic sequence converges.

Coming up next, we're going to return to the definitions and theorems of subsequences from Chapter 15 and look at their limits in more detail.

CHAPTER 17

Subsequential Limits

Recall our classic example of the "alternating" sequence, $p_n = (-1)^n + \frac{(-1)^n}{n}$ for every $n \in \mathbb{N}$ (refer back to Figures 14.4 and 14.5). This sequence diverges, even though it looks like it's really converging to two different limits. To better understand these types of sequences, we should develop a theory for sequences that may diverge but have such prominent subsequential limits.

In Theorem 15.9, we took a sequence $\{p_n\}$ and we looked at the set E^*, which was composed of every one of its subsequential limits (meaning, the limit of every one of its convergent subsequences). Our goal, then, is to study this type of set in more detail, specifically investigating the properties of its upper and lower bounds. Why? Because knowing about these bounds can sometimes give us valuable information about sequences like $p_n = (-1)^n + \frac{(-1)^n}{n}$.

To look at the bounds of the set of all subsequential limits of a sequence, it's not enough to know about which subsequences converge. We also want to know if any divergent subsequences become infinitely large or small. For example, take the sequence $1, 2, 1, 3, 1, 4, 1, 5, \ldots$. The only subsequential limit here is 1 (since $1, 1, 1, \ldots$ is a subsequence), but there are other subsequences (such as $2, 3, 4, \ldots$) that increase to infinity. So it's not really fair to say that the upper bound of all subsequential limits is 1; really, it has no upper bound, since many subsequences get arbitrarily large.

Given all this hoopla, what we need is a way to distinguish between sequences that diverge and sequences that diverge but get arbitrarily large. That's why, for the rest of this chapter, we'll work with sequences in the extended real number system $\mathbb{R} \cup \{+\infty, -\infty\}$, which is explained in Definition 5.10.

Don't panic! At least this will be easier than working in $\mathbb{R}^k$, or an arbitrary metric space. And everything we do should also work in any ordered field that has the least upper bound property and also has a meaningful definition of $+\infty$ and $-\infty$.

Definition 17.1. *(Diverging to Infinity)*
For any sequence $\{s_n\}$ in the metric space $\mathbb{R}$, $\{s_n\}$ **diverges to infinity** if for every $M \in \mathbb{R}$, $s_n \geq M$ for every n greater than or equal to some natural number N.

In symbols, we write $\lim_{n \to \infty} s_n = +\infty$ (or $s_n \to +\infty$ for short) if:

$$\forall M \in \mathbb{R}, \exists N \in \mathbb{N} \text{ such that } n \geq N \implies s_n \geq M.$$

Similarly, $\{s_n\}$ also diverges to infinity if for every $M \in \mathbb{R}$, $s_n \leq M$ for every n greater than or equal to some natural number N.

In symbols, we write $\lim_{n \to \infty} s_n = -\infty$ (or $s_n \to -\infty$ for short) if:

$$\forall M \in \mathbb{R}, \exists N \in \mathbb{N} \text{ such that } n \geq N \implies s_n \leq M.$$

Using the limit notation and the arrow symbol ($\to$) for sequences that diverge to infinity is an egregious abuse of notation. We are *not* saying that $\{s_n\}$ converges in any way. Rather, we are saying that $\{s_n\}$ diverges and gets arbitrarily large. The only reason we use the same symbols as for convergent sequences is that it will be more convenient when defining the set of all subsequential limits (for the set to possibly include $+\infty$ and $-\infty$).

Example 17.2. (Diverging to Infinity)
The sequence given by $s_n = n^2$ for every $n \in \mathbb{N}$ diverges to infinity, so we write $s_n \to \infty$. If $s_n = -5n$ for every $n \in \mathbb{N}$, then $s_n \to -\infty$. The sequence given by $s_n = (-1)^n$ for every $n \in \mathbb{N}$ diverges but does not diverge to infinity.

Similarly, if $s_n = (-1)^n n$ for every $n \in \mathbb{N}$, then $\{s_n\}$ diverges, but does not diverge to infinity. Why? Don't the values of $\{s_n\}$ get arbitrarily close to $+\infty$ and $-\infty$? Yes, but that is precisely the problem! The sequence fluctuates between large positive numbers and large negative numbers. Given any $M \in \mathbb{R}$, we cannot say $s_n \geq M$ for *all* n greater than or equal to some $N \in \mathbb{N}$, since $s_{n+1} = -(s_n + 1) < M$. The same problem occurs if we want $s_n \leq M$. On the other hand, $\{s_n\}$ does have subsequences that diverge to $+\infty$, and others that diverge to $-\infty$.

Theorem 17.3. *(Unbounded $\iff$ A Subsequence Diverges to Infinity)*
For any sequence $\{s_n\}$ in the metric space $\mathbb{R}$, $\{s_n\}$ is unbounded if and only if some subsequence of $\{s_n\}$ diverges to infinity.

Proof. If $\{s_n\}$ is unbounded, then for every $q \in \mathbb{R}$ and every $M \in \mathbb{R}$, there is an element s_n of the sequence with $|s_n - q| > M$. So for any M there is an element of $\{s_n\}$ with $s_n \geq M$ (or $s_n \leq M$). Because M is an arbitrary number, there is also an element of $\{s_n\}$ with $s_n \geq M + 1$ (or $s_n \leq M - 1$), another with $s_n \geq M + 2$ (or $s_n \leq M - 2$), and so on. Therefore, there are infinitely many elements of $\{s_n\}$ that are greater (or less than) M, so the subsequence composed of these elements diverges to infinity.

If some subsequence $\{s_{n_k}\}$ of $\{s_n\}$ diverges to infinity, then for any $M \in \mathbb{R}$ we can find an $N \in \mathbb{N}$ such that $s_{n_k} \geq M$ whenever $k \geq N$ (we're assuming for simplicity that it diverges to positive infinity, since the same argument works if it were negative infinity). So given any $q \subset \mathbb{R}$ and any $M < \infty$, there are infinitely many elements of $\{s_n\}$ with $s_n \geq M + q + 1$, meaning $s_n - q < M$; similarly, there are infinitely many elements of $\{s_n\}$ with $s_n \geq -M + q + 1$, meaning $s_n - q > -M$. Thus $|s_n - q| > M$ for at least one element of $\{s_n\}$, so $\{s_n\}$ is not bounded. $\square$

The following is the major definition we have been working toward.

Definition 17.4. *(Upper and Lower Limits)*
For any sequence $\{s_n\}$ in the metric space $\mathbb{R}$, let E be the set of all numbers $x \in \mathbb{R} \cup \{+\infty, -\infty\}$ such that $s_{n_k} \to x$ for some subsequence $\{s_{n_k}\}$.

Let $s^* = \sup E$, and let $s_* = \inf E$. Then s^* is the **upper limit** of $\{s_n\}$, and s_* is the **lower limit** of $\{s_n\}$; we write $\limsup_{n \to \infty} s_n = s^*$, and $\liminf_{n \to \infty} s_n = s_*$.

 When we dealt with the set of all subsequential limits in Theorem 15.9, we called that set E^*. Here, the set E is slightly different—it is the set of all subsequential limits of $\{s_n\}$, plus possibly $+\infty$ and/or $-\infty$ if any subsequence $\{s_{n_k}\}$ of $\{s_n\}$ diverges to infinity. Our definition allows this, since if x is in the extended real number system and $s_{n_k} \to x$, then x can be a real number, or it can be $+\infty$ or $-\infty$ if the subsequence $\{s_{n_k}\}$ diverges to infinity.

Remember that in the extended real number system, if a set is not bounded above, then its supremum is $+\infty$, and if a set is not bounded below, then its infimum is $-\infty$. So if there is a subsequence $\{s_{n_k}\}$ of $\{s_n\}$ with $s_{n_k} \to +\infty$, then $s^* = +\infty$; and if $s_{n_k} \to -\infty$, then $s_* = -\infty$.

For convenience, in the rest of this chapter, instead of writing "let E be the set of all numbers $x \in \mathbb{R} \cup \{+\infty, -\infty\}$ such that $s_{n_k} \to x$ for some subsequence $\{s_{n_k}\}$," we will just write "let E be the set of all subsequential limits* of $\{s_n\}$." The star (*) is there to indicate that E is the set of all subsequential limits, *including* possibly $+\infty$ and/or $-\infty$ if any subsequence of $\{s_n\}$ diverges to infinity. (Always remember, though, that $+\infty$ and $-\infty$ are *not* actually limits.)

Rant. I hate the notation "lim sup." It is horribly confusing! The upper limit is *not* the limit of some kind of sequence of suprema, as the symbols might lead you to believe. Really, the upper limit is the supremum of all subsequential limits. So it should be written more like "sup lim" to help you remember.

Also, the upper limit is not a limit. It is a bound on the set of *subsequential* limits. There's nothing in the notation that clarifies that subsequences are involved!

The "$n \to \infty$" makes it even worse. With sequences, the notations $\lim_{n \to \infty} s_n = s$ and $s_n \to s$ mean the same thing. So you might think that we could write $\limsup_{n \to \infty} s_n = s^*$ as $\sup s_n \to s^*$. But we can't. Writing $\sup s_n \to s^*$ is equivalent to writing $\lim_{n \to \infty} \sup s_n = s^*$ (notice that here, the "$n \to \infty$" is underneath the "lim," not the "sup"). Either way, that wouldn't make much sense, since each s_n is a single point, not a set, and you can only take the supremum of a set.

Worst of all, most people pronounce "lim sup" as "limb soup," which doesn't sound particularly appetizing.

We have used a common definition of lim sup, but you should be aware that other (equivalent) definitions exist. One other definition does define the lim sup as the limit of a sequence of suprema, which is more consistent with the notation—but that definition is more difficult to work with when proving the upcoming theorems.

Example 17.5. (Upper and Lower Limits)
For each of the following sequences in the metric space $\mathbb{R}$, let E be the set of all its subsequential limits*.

1. If $s_n = [(-1)^n + 1]n$ for every $n \in \mathbb{N}$, then as we can see in Figure 17.1, $s_n = 0$ for odd n, and $s_n = 2n$ for even n. Then every subsequence of $\{s_n\}$ either converges to 0 or diverges to infinity. (Remember, a number like 12 is not a subsequential limit, because any subsequence that starts out like 4, 8, 12 must keep going for an infinite number of steps.)

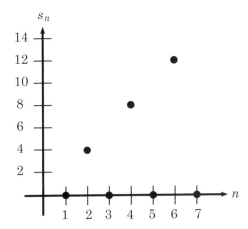

Figure 17.1. The first few elements of the sequence $s_n = [(-1)^n + 1]n$.

Let's take $n_k = 2k$ for every $k \in \mathbb{N}$, so the subsequence $\{s_{n_k}\}$ is $s_2, s_4, s_6, \ldots = 4, 8, 12, \ldots$ In other words, $s_{n_k} = 4k$ for every $k \in \mathbb{N}$, and we want to prove it diverges to infinity. Given any $M \in \mathbb{R}$, we need an $N \in \mathbb{N}$ such that $s_{n_k} \geq M$ whenever $k \geq N$. Just let $N = \lceil \frac{M}{4} \rceil$, so that

$$k \geq N \implies s_{n_k} = 4k \geq 4 \lceil \tfrac{M}{4} \rceil \geq M.$$

Thus the set of subsequential limits* of $\{s_n\}$ is the two-element set $E = \{0, +\infty\}$. Then $s^* = +\infty$ and $s_* = 0$, or in other words

$$\limsup_{n \to \infty} s_n = +\infty, \text{ and } \liminf_{n \to \infty} s_n = 0.$$

2. If $s_n = 1$ for every $n \in \mathbb{N}$, then every subsequence of $\{s_n\}$ converges to 1. Thus $E = \{1\}$, so $s^* = 1$ and $s_* = 1$, or in other words

$$\limsup_{n \to \infty} s_n = 1, \text{ and } \liminf_{n \to \infty} s_n = 1.$$

3. Recall from Example 14.2 that the elements of $\mathbb{Q}$ can be arranged into some sequence $\{s_n\}$. Given any real number x, we can actually build a subsequence of $\{s_n\}$ that converges to x. Just take an increasingly accurate decimal expansion of x, as in Example 16.2.

    Wait, how? We never saw the explicit mapping from $\mathbb{N}$ to $\mathbb{Q}$, so we don't know anything about the order of the elements in $\{s_n\}$. If we are trying to find a subsequence that converges to $\sqrt{2}$, we can't necessarily take $1.4, 1.41, 1.414, \ldots$, since 1.41 might come before 1.4 in the sequence $\{s_n\}$. On the other hand, we *do* know that there are an infinite amount of rational numbers close to $\sqrt{2}$. So find 1.4 in the sequence, and look for 1.41. If 1.41 has already occurred in $\{s_n\}$, look for 1.414; if 1.414 has already occurred, look for 1.4142, and so on. Because there are an infinite number of points leading up to $\sqrt{2}$ and only finitely many of them could have occurred in $\{s_n\}$ before 1.4, there must be infinitely many after 1.4, so the subsequence will be valid.

By the same logic, there are subsequences of $\{s_n\}$ that diverge to both positive and negative infinity, since the range of $\{s_n\}$ is $\mathbb{Q}$, which is bounded neither above nor below. Thus E is actually the entire extended real number system, so $s^* = +\infty$ and $s_* = -\infty$, or in other words,

$$\limsup_{n \to \infty} s_n = +\infty, \text{ and } \liminf_{n \to \infty} s_n = -\infty.$$

4. Take our classic example of the "alternating" sequence, $p_n = (-1)^n + \frac{(-1)^n}{n}$ for every $n \in \mathbb{N}$ (see Figures 14.4 and 14.5). The sequence diverges, but the even elements converge to 1 and the odd elements converge to -1. Note that no subsequence can converge to anything else.

Thus $E = \{-1, 1\}$, so $s^* = 1$ and $s_* = -1$, or in other words

$$\limsup_{n \to \infty} s_n = 1, \text{ and } \liminf_{n \to \infty} s_n = -1.$$

This is why looking at upper and lower limits can be useful; if all we say is "this sequence diverges," we completely lose out on the fact that it really looks like it's converging to two different points.

Theorem 17.6. *(Upper and Lower Limits of Convergent Sequences)*
For any sequence $\{s_n\}$ in the metric space $\mathbb{R}$, $\{s_n\}$ converges to the finite number $s \in \mathbb{R}$ if and only if

$$\limsup_{n \to \infty} s_n = \liminf_{n \to \infty} s_n = s.$$

Proof. To prove the second direction, we will show that every subsequence of $\{s_n\}$ is bounded. Then we can apply the Bolzano-Weierstrass theorem (Theorem 15.8) to see that some subsequence of any subsequence of $\{s_n\}$ must converge to s. If $\{s_n\}$ did not converge to s, some subsequence would not converge to s, which would give us a contradiction.

To do this formally, we assume $\limsup_{n \to \infty} s_n = s$ and $\liminf_{n \to \infty} s_n = s$ for the same finite number $s \in \mathbb{R}$. Then the set E of every subsequential limit* of $\{s_n\}$ consists of the single point s. So every convergent subsequence of $\{s_n\}$ converges to s.

Also, no subsequence of $\{s_n\}$ diverges to infinity (because otherwise $\limsup_{n \to \infty} s_n$ would be $+\infty$, or $\liminf_{n \to \infty} s_n$ would be $-\infty$). So by Theorem 17.3, $\{s_n\}$ is bounded.

Now if the sequence $\{s_n\}$ does *not* converge to s, then there exists an $\epsilon > 0$ such that for infinitely many natural numbers n, we have $s_n - s \geq \epsilon$ (or $s - s_n \geq \epsilon$). Let $\{s_{n_k}\}$ be the subsequence consisting of all such elements of $\{s_n\}$. Because $\{s_n\}$ is bounded, so is $\{s_{n_k}\}$, so by the Bolzano-Weierstrass theorem, some subsequence $\{s_{n_{k_j}}\}$ of $\{s_{n_k}\}$ converges. But every element of $\{s_{n_k}\}$ is $\geq s + \epsilon$ (or $\leq s - \epsilon$) so every element of $\{s_{n_{k_j}}\}$ is $\geq s + \epsilon$ (or $\leq s - \epsilon$), so $\{s_{n_{k_j}}\}$ cannot converge to s. (Such convergence is impossible because ϵ is a fixed positive number. Thus we have a subsequence of $\{s_n\}$ which converges to something other than s, which is a contradiction. So $\{s_n\}$ must converge to s.

The other direction is much easier. Assume $\{s_n\}$ converges to some point $s \in \mathbb{R}$. Then by Theorem 15.6, every subsequence of $\{s_n\}$ converges to s. Now the set E of every subsequential limit* of $\{s_n\}$ consists of the single point s, so

$$\limsup_{n \to \infty} s_n = \sup E = \sup\{s\} = s = \inf\{s\} = \sup E = \liminf_{n \to \infty} s_n.$$

□

The next question we might ask is, "are the upper and lower limits of any sequence necessarily limits of some subsequence?" We have seen examples of sets that do not contain their suprema and infima. So does the set of all subsequential limits* of any sequence contain its supremum and infimum? The answer is ... drum roll, please ... Yes!

Theorem 17.7. *(Upper and Lower Limits Are Subsequential Limits*)*
For any sequence $\{s_n\}$ in the metric space $\mathbb{R}$, let E be the set of its subsequential limits*. Then $s^* = \limsup_{n \to \infty} s_n$ is an element of E, and so is $s_* = \liminf_{n \to \infty} s_n$.

In other words, there is some subsequence of $\{s_n\}$ that converges to s^*, and there is some subsequence of $\{s_n\}$ that converges to s_*.

Proof. Let's work with the lim sup first. There are three possible cases.

Case 1. $s^* = +\infty$. Then E is not bounded above in $\mathbb{R}$, so given any $N \in \mathbb{R}$, there is a subsequence $\{s_{n_k}\}$ of $\{s_n\}$ that converges to something $\geq N$. So for any $\epsilon > 0$, there are infinitely elements of $\{s_{n_k}\}$ which are $\geq N - \epsilon$. Then fix ϵ and let $M = N - \epsilon$, and note that every element of $\{s_{n_k}\}$ is also an element of $\{s_n\}$. So given any number $M \in \mathbb{R}$, there are infinitely elements of $\{s_n\}$ that are $\geq M$, so there is some subsequence that diverges to $+\infty$. Thus $+\infty \in E$, so $s^* \in E$.

Case 2. $s^* \in \mathbb{R}$. Then E is bounded above, and remember from Theorem 15.9 that E is closed. By Corollary 10.11, closed sets that are bounded above contain their suprema, so $s^* \in E$.

Case 3. $s^* = -\infty$. Then no element of E is greater than $-\infty$, so $-\infty$ must be the only element of E. So $s^* \in E$.

The proof for the lim inf is basically the same. Fill in the blanks in Box 17.1, *por favor!*

BOX 17.1

PROVING THEOREM 17.7 FOR THE lim inf
Case 1. $s_* = +\infty$. Then no element of E is _____ than $+\infty$, so $-\infty$ must be the only element of E. So $s_* \in$ _____ .
Case 2. $s_* \in \mathbb{R}$. Then E is bounded _____ , and remember from Theorem _____ that E is closed. By Corollary 10.11, _____ sets that are bounded below contain their infima, so _____ $\in E$.

Subsequential Limits • 163

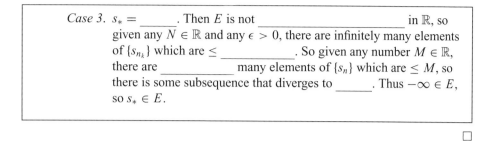

Case 3. $s_* = $ _____. Then E is not _____ in $\mathbb{R}$, so given any $N \in \mathbb{R}$ and any $\epsilon > 0$, there are infinitely many elements of $\{s_{n_k}\}$ which are $\leq$ _____. So given any number $M \in \mathbb{R}$, there are _____ many elements of $\{s_n\}$ which are $\leq M$, so there is some subsequence that diverges to _____. Thus $-\infty \in E$, so $s_* \in E$. $\square$

We might also want to know if pinpointing the upper limit of a sequence tells us anything about the sequence itself, not just its subsequences. It turns out that the upper limit is actually an upper bound on every element of the sequence (past a certain step).

Theorem 17.8. *(Upper and Lower Limits as Bounds of the Sequence)*
For any sequence $\{s_n\}$ in the metric space $\mathbb{R}$, let E be the set of its subsequential limits*, and let $s^* = \limsup_{n\to\infty} s_n$. Then for any $x > s^*$, there is an $N \in \mathbb{N}$ such that $s_n < x$ whenever $n \geq N$.

Similarly, let $s_* = \liminf_{n\to\infty} s_n$. Then for any $x < s_*$, there is an $N \in \mathbb{N}$ such that $s_n > x$ whenever $n \geq N$.

In other words, any number greater than the upper limit is also greater than any element of the sequence, past some step N.

Notice that we can only take an x with $x > s^*$ if s^* is not $+\infty$. Similarly, we cannot take an x with $x < s_*$ if $s_* = -\infty$.

Proof. Let's do a proof by contradiction. If there is an $x > s^*$ such that given any $N \in \mathbb{N}$, $s_n \geq x$ for some $n \geq N$, then $s_n \geq x$ for infinitely many values of n. We can form a subsequence $\{s_{n_k}\}$ with all of those elements of $\{s_n\}$, where every element of the subsequence is $s_{n_k} \geq x$.

Now we make the same argument as in the proof of Theorem 17.6.

Case 1. If $\{s_{n_k}\}$ is not bounded, then by Theorem 17.3, it has a subsequence $\{s_{n_{k_j}}\}$ that diverges to infinity. Then $+\infty \in E$ (we cannot have $s_{n_{k_j}} \to -\infty$, because we know every element of $\{s_{n_k}\}$ is greater than a fixed number x). This is a contradiction, because $+\infty > x > s^*$, but s^* is supposed to be the supremum of E.

Case 2. If $\{s_{n_k}\}$ is bounded, then by the Bolzano-Weierstrass theorem, it has a subsequence $\{s_{n_{k_j}}\}$ that converges. Because every element of $\{s_{n_k}\}$ is $\geq x$, so is every element of $\{s_{n_{k_j}}\}$, so this subsequence must converge to some point $y \geq x$, and $y \in E$. This is a contradiction, because $y \geq x > s^*$, but s^* is supposed to be the supremum of E.

Note how we made use of the seemingly unnecessary value x. The theorem would not necessarily be true if it stated "there is an $N \in \mathbb{N}$ such that $s_n \leq s^*$ whenever $n \geq N$," since we might only be able to find a subsequence that converges to $y \geq s^*$. This would not give us a contradiction, because y could $= s^*$, which does not disagree with the fact that $s^* = \sup E$. We needed an x strictly greater than s^* to end up with $y > s^*$.

The proof for s_* is basically the same. Fill in the blanks in Box 17.2, *s'il vous plaît!*

BOX 17.2

> PROVING THEOREM 17.8 FOR THE lim inf
>
> If there is an $x < s_*$ such that given any $N \in \mathbb{N}$, $s_n \leq x$ for some $n \geq N$, then we can form a subsequence $\{s_{n_k}\}$ where every element of the subsequence is $s_{n_k} \leq$ _____.
>
>   *Case 1.* If $\{s_{n_k}\}$ is not bounded, then by Theorem _____, it has a subsequence $\{s_{n_{k_j}}\}$ that _____ to infinity. Then _____ $\in E$, which is a contradiction because _____ $< x < s_*$.
>
>   *Case 2.* If $\{s_{n_k}\}$ is bounded, then by the _____ theorem, it has a subsequence $\{s_{n_{k_j}}\}$ which _____. Because every element $\{s_{n_k}\}$ is $\leq x$, so is every element of _____, so this subsequence must converge to some point $y \leq x$. Then _____ $\in E$, which is a contradiction because _____ $\leq x < s_*$.

□

Those previous few theorems help us prove that for any sequence, there is always exactly one upper limit and exactly one lower limit. This is equivalent to asserting both existence (meaning, there is at least one upper limit and at least one lower limit) and uniqueness (meaning, there is at most one upper limit and at most one lower limit).

Theorem 17.9. *(Existence and Uniqueness of Upper and Lower Limits)*
For any sequence $\{s_n\}$ in the metric space $\mathbb{R}$, $s^* = \limsup_{n \to \infty} s_n$ exists (in the extended real number system) and is unique, and $s_* = \limsup_{n \to \infty} s_n$ exists (in the extended real number system) and is unique.

Proof. To prove existence, we just need to show that the set of subsequential limits* E is not empty. Because then either E is not bounded above, in which case $s^* = +\infty$, or else E is bounded above, and we can use the least upper bound property of $\mathbb{R}$ to assert the existence of $s^* = \sup E$. Similarly, if E is not bounded below, $s^* = -\infty$; otherwise, we can use the greatest lower bound property of $\mathbb{R}$ to assert the existence of $s_* = \sup E$.

To show that E is nonempty, we'll use our classic argument. If $\{s_n\}$ is unbounded, then by Theorem 17.3, some sequence diverges to infinity, so either $+\infty \in E$ or $-\infty \in E$. Otherwise, $\{s_n\}$ is bounded, so by the Bolzano-Weierstrass theorem, there is a subsequence that converges to some point s, so $s \in E$.

To prove uniqueness, let's start with the lim sup. If two different numbers p and q exist with $p < q$, and both $p = \limsup_{n \to \infty} s_n$ and $q = \limsup_{n \to \infty} s_n$, then we have a contradiction. Why? Take any real number x with $p < x < q$. Then by Theorem 17.8, there exists an $N \in \mathbb{N}$ such that $s_n < x$ whenever $n \geq N$. But then every subsequence of $\{s_n\}$ can only converge to a number $\leq x$, so no subsequence of $\{s_n\}$ can converge to q (since $q > x$). Thus $q \notin E$, which contradicts Theorem 17.7.

Similarly, if two different numbers p and q exist with $p < q$, and both $p = \liminf_{n \to \infty} s_n$ and $q = \liminf_{n \to \infty} s_n$, then take $p < x < q$. By Theorem 17.8, there exists an $N \in \mathbb{N}$ such that $s_n > x$ whenever $n \geq N$. But then no subsequence of $\{s_n\}$ can converge to p, which contradicts Theorem 17.7. □

Theorem 17.10. *(Comparing Upper and Lower Limits)*
For any sequences $\{s_n\}$ and $\{t_n\}$ in the metric space $\mathbb{R}$, let N be any natural number. Then if $s_n \leq t_n$ for every $n \geq N$, the upper limit of $\{t_n\}$ is greater than or equal to the upper limit of $\{s_n\}$, and the lower limit of $\{t_n\}$ is greater than or equal to the lower limit of $\{s_n\}$.

In symbols:

$$\forall N \in \mathbb{N}, \textit{ if } s_n \leq t_n \forall n \geq N, \textit{ then}$$

$$\limsup_{n \to \infty} s_n \leq \limsup_{n \to \infty} t_n,$$

$$\liminf_{n \to \infty} s_n \leq \liminf_{n \to \infty} t_n.$$

Proof. Let E be the set of subsequential limits* of $\{s_n\}$, and let F be the set of subsequential limits* of $\{t_n\}$. Fix an $N \in \mathbb{N}$, and take any sequence $\{n_k\}$. Then for infinitely many k, we have $s_{n_k} \leq t_{n_k}$. So if $s_{n_k} \to s$ and $t_{n_k} \to t$, then $s \leq t$; if $s_{n_k} \to +\infty$, then clearly also $t_{n_k} \to +\infty$; if $t_{n_k} \to -\infty$, then clearly also $s_{n_k} \to -\infty$.

Because this is true for every possible sequence $\{n_k\}$, we see that every element of E is less than or equal to the corresponding element of F. Hence $\sup E \leq \sup F$, and $\inf E \leq \inf F$. □

Whew! Those were a lot of theorems about upper and lower limits. Remember the main techniques we used to investigate the set of all subsequential limits* E of a real sequence $\{s_n\}$—they will come in handy in the future:

1. If anything is true for an infinite number of elements of $\{s_n\}$, then we can make a subsequence out of those elements.
2. If $\{s_n\}$ (or any subsequence of $\{s_n\}$) is unbounded, then by Theorem 17.3, it has one or more subsequences that diverge, to $+\infty \in E$ and/or $-\infty \in E$.
3. If $\{s_n\}$ (or any subsequence of $\{s_n\}$) is bounded, then by the Bolzano-Weierstrass theorem, it has a subsequence that converges to some point $s \in \mathbb{R}$, so $s \in E$.

CHAPTER 18

Special Sequences

Before we conclude our study of sequences and move on to series, let's take a look at some important sequences in $\mathbb{R}$—which come up over and over again in the study of real analysis—and prove that they converge. These sequences (and their limits) are:

1. $\frac{1}{n^p} \to 0$ (if $p > 0$).

2. $\sqrt[n]{p} \to 1$ (if $p > 0$).

3. $\sqrt[n]{n} \to 1$.

4. $\frac{n^\alpha}{(1+p)^n} \to 0$ (if $p > 0$ and $\alpha \in \mathbb{R}$).

5. $x^n \to 0$ (if $|x| < 1$).

To prove their convergence, it will be helpful to first prove what's known as the pinching theorem for sequences of real numbers.

Theorem 18.1. *(The Pinching Theorem)*
Let $\{s_n\}$, $\{a_n\}$, and $\{b_n\}$ be sequences in the metric space $\mathbb{R}$ with $a_n \leq s_n \leq b_n$ for every $n \in \mathbb{N}$. If a_n and b_n converge to the same real number s, then so does s_n.

In symbols, for any sequences $\{s_n\}$, $\{a_n\}$, and $\{b_n\}$ in $\mathbb{R}$:

$$a_n \leq s_n \leq b_n \ \forall n \in \mathbb{N}, \ \lim_{n \to \infty} a_n = s \text{ and } \lim_{n \to \infty} b_n = s \implies \lim_{n \to \infty} s_n = s.$$

Basically, as we can see in Figure 18.1, if any sequence $\{s_n\}$ can be squeezed between two other sequences that converge to the same point, then as n goes to infinity, s_n is "pinched" to that point.

Proof. For any $\epsilon > 0$, we can apply the definition of convergence to obtain two natural numbers N_1 and N_2 such that

$$n \geq N_1 \implies d(a_n, s) < \epsilon,$$

$$n \geq N_2 \implies d(b_n, s) < \epsilon.$$

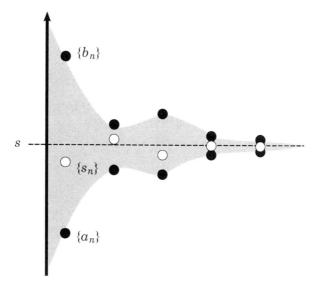

Figure 18.1. Any sequence between two other sequences that converge to the same point s will also be "pinched" to the limit s.

Let $N = \max\{N_1, N_2\}$, so for $n \geq N$, we have

$$s_n - s \leq b_n - s \leq |b_n - s| < \epsilon,$$

and

$$-s_n + s \leq -a_n + s \leq |a_n - s| < \epsilon.$$

Thus

$$|s_n - s| = \max\{s_n - s, -(s_n - s)\} < \epsilon.$$

Since this is true for every $\epsilon > 0$, we have $s_n \to s$. $\square$

As promised, the rest of the chapter is devoted to our super-duper special sequences.

Theorem 18.2. *(The Sequence n^p)*
If $p > 0$, then $\lim_{n \to \infty} \frac{1}{n^p} = 0$. (In other words, if $p < 0$, then $\lim_{n \to \infty} n^p = 0$.)

Proof. We want to find an $N \in \mathbb{N}$ such that

$$n \geq N \implies d\left(\frac{1}{n^p}, 0\right) < \epsilon.$$

So we need $n^p \epsilon > 1$. Well $\sqrt[p]{\frac{1}{\epsilon}}$ exists in $\mathbb{R}$ by Theorem 5.8, so just let

$$N = \left\lceil \frac{1}{\sqrt[p]{\epsilon}} \right\rceil + 1.$$

Of course, we haven't explicitly proved the theorem according to the definition of convergence; we just took care of the hard work of finding an N that will work. But after so many convergence proofs, hopefully you have the formal part down cold! □

Theorem 18.3. *(The Sequence $\sqrt[n]{p}$)*
If $p > 0$, then $\lim_{n \to \infty} \sqrt[n]{p} = 1$.

Proof. There are three possible cases.

Case 1. $p > 1$. Let $x_n = \sqrt[n]{p} - 1$, so that $x_n > 0$. The goal is to prove that every element of $\{x_n\}$ is less than some other sequence s_n that converges to 0. Then we'll have $0 \leq x_n \leq s_n$, so by the pinching theorem, $\lim_{n \to \infty} x_n = 0$ (since the sequence $0, 0, 0, \ldots$ converges to 0). By Theorem 15.1 we can add a constant to the inside of a limit, so we will have

$$1 = 1 + \lim_{n \to \infty} x_n = \lim_{n \to \infty} (x_n + 1) = \lim_{n \to \infty} \sqrt[n]{p}.$$

How do we choose such a sequence $\{s_n\}$ that works? This step requires some creativity, so let's think it through. We know that the sequence $\frac{1}{n}$ converges to 0. On the other hand, we can't be sure that $x_n \leq \frac{1}{n}$ for every $n \in \mathbb{N}$. We know that p can be written in terms of x_n, so we can try to incorporate p into $\frac{1}{n}$ to somehow make it greater than x_n. Since p is a constant, remember from Theorem 15.1 that

$$\lim_{n \to \infty} \frac{p}{n} = p \lim_{n \to \infty} \frac{1}{n} = 0.$$

How can we show that $\frac{p}{n} > x_n$? We'll use the binomial theorem (derived from basic algebra) which tells us how to expand powers of the form $(a + b)^n$:

$$(a + b)^n = \sum_{k=0}^{n} \binom{n}{k} a^{n-k} b^k$$

$$= \sum_{k=0}^{n} \frac{n!}{(n-k)!k!} a^{n-k} b^k$$

$$= a^n + na^{n-1}b + \frac{n(n-1)}{2} a^{n-2} b^2 + \ldots$$

$$+ \frac{n(n-1)}{2} a^2 b^{n-2} + nab^{n-1} + b^n.$$

As you can see in the equation, the symbol $\binom{n}{k}$ just represents $\frac{n!}{(n-k)!k!}$. It is read "n choose k," and appears frequently in the study of probability. And in case you aren't familiar with the $n!$ symbol either, it means multiply n by each consecutive natural number lower than it; meaning, $n! = n(n-1)(n-2)\dots 1$. That expression is read as "n factorial," though I suppose you could instead shout "n!" loudly.

We calculate

$$p = (1 + x_n)^n$$

$$= \sum_{k=0}^{n} \binom{n}{k} 1^{n-k} x_n^k \quad \text{(by the binomial theorem)}$$

$$= \left(1\right)\left(x_n^0\right) + \left(n\right)\left(x_n^1\right) + \left(\tfrac{n(n-1)}{2}\right)\left(x_n^2\right) + \dots$$

$$+ \left(\tfrac{n(n-1)}{2}\right)\left(x_n^{n-2}\right) + \left(n\right)\left(x_n^{n-1}\right) + \left(1\right)\left(x_n^n\right)$$

$$> 1 + nx_n.$$

In that last step, we got rid of every term beyond the first two, since $x_n > 0$ means that they are all positive. Then $0 < x_n < \tfrac{p-1}{n}$. (Notice we didn't quite get $\tfrac{p}{n} > x_n$, since there's that extra 1 tacked on, but it really doesn't matter, because $p - 1$ is still just a constant.)

So by the pinching theorem, $x_n \to 0$ (notice that the pinching theorem still works when we replace $\leq$ with $<$). Thus

$$\lim_{n \to \infty} \sqrt[n]{p} = \lim_{n \to \infty} (x_n + 1) = 1 + \lim_{n \to \infty} x_n = 1 + 0 = 1.$$

Case 2. $p = 1$. We have $\lim_{n \to \infty} \sqrt[n]{p} = \lim_{n \to \infty} 1 = 1$.

Case 3. $0 < p < 1$. We can use the same argument from the first case, but with the inequalities reversed. Try filling in the blanks in Box 18.1.

BOX 18.1

PROVING $\sqrt[n]{p} \to 0$ WHEN $0 < p < 1$

Let $x_n = $ _____, so that $x_n < 0$. Then

$$p = (1 + x_n)^n$$

$$= \sum_{k=0}^{n} \binom{n}{k} 1^{n-k} \underline{\qquad} \quad \text{(by the binomial theorem)}$$

$$= \left(1\right)\left(x_n^0\right) + \left(n\right)\left(x_n^1\right) + \left(\tfrac{n(n-1)}{2}\right)\left(x_n^2\right) + \dots$$

$$+ \underline{\hspace{2cm}} + \underline{\hspace{2cm}} + \underline{\hspace{2cm}}$$

$$< 1 + nx_n \quad \text{(since } x_n < 0 \text{ means every term beyond the first two is } < 0\text{).}$$

Then $\tfrac{p-1}{n} < x_n < 0$, so by the _____, $x_n \to 0$.
Thus $\lim_{n \to \infty} \sqrt[n]{p} = 1$.

Theorem 18.4. *(The Sequence $\sqrt[n]{n}$)*
$\lim_{n \to \infty} \sqrt[n]{n} = 1$.

How is this different from the previous theorem? Back there, we were looking at the sequence $\{\sqrt[n]{p}\}$ for a fixed number $p > 0$. Here, the number we are nth-rooting is n itself. So this sequence is:

$$\{\sqrt[n]{n}\} = 1, \sqrt{2}, \sqrt[3]{3}, \sqrt[4]{4}, \sqrt[5]{5}, \sqrt[6]{6}, \sqrt[7]{7}, \ldots$$

If we look at the sequence in decimal form (to two decimal places), we can tell that it begins at 1, jumps up above 1.4, then starts decreasing closer and closer back to 1:

$$\{\sqrt[n]{n}\} \approx 1.00, 1.41, 1.44, 1.41, 1.38, 1.35, \ldots$$

It isn't immediately obvious that it converges to 1, because it isn't decreasing very fast. That's why knowing how to prove these things comes in handy!

Proof. We can use the same trick as before: let $x_n = \sqrt[n]{n} - 1$. This time, it will be even easier, since $x_n \geq 0$ for every $n \in \mathbb{N}$ means we only have to worry about one case.

We want to squeeze x_n below a sequence that converges to 0, but note that if we use the same one we found in the previous theorem, $\frac{p-1}{n}$, here it would be $\frac{n-1}{n}$. We don't know if this sequence converges to 0 (in fact, it doesn't!) since the top is not constant. Let's apply the binomial theorem as before, but this time look for a term with an extra n, so it can hopefully "cancel out" that n at the bottom of the fraction.

$$n = (1 + x_n)^n$$
$$= \sum_{k=0}^{n} \binom{n}{k} 1^{n-k} x_n^k \quad \text{(by the binomial theorem)}$$
$$= (1)(x_n^0) + (n)(x_n^1) + \left(\frac{n(n-1)}{2}\right)(x_n^2) + \ldots$$
$$+ \left(\frac{n(n-1)}{2}\right)(x_n^{n-2}) + (n)(x_n^{n-1}) + (1)(x_n^n)$$
$$\geq \left(\frac{n(n-1)}{2}\right)(x_n^2) \quad \text{(since } x_n \geq 0 \text{ means every other term is } \geq 0\text{)}.$$

Then

$$0 \leq x_n \leq \sqrt{\frac{n}{\frac{n(n-1)}{2}}} = \sqrt{\frac{2}{n-1}}.$$

If we can show that $\sqrt{\frac{2}{n-1}} \to 0$, we can use the pinching theorem to see that $x_n \to 0$, so that $\sqrt[n]{n} \to 1$. To do so, we simply apply Theorem 18.2 with $p = \frac{1}{2}$ to see that $\frac{1}{\sqrt{n}} \to 0$. Then clearly $\frac{1}{\sqrt{n-1}} \to 0$, so indeed $\sqrt{2} \frac{1}{\sqrt{n-1}} \to 0$. □

Theorem 18.5. *(The Sequence $n^\alpha (1+p)^{-n}$)*
If $p > 0$ and $\alpha \in \mathbb{R}$, then $\lim_{n \to \infty} \frac{n^\alpha}{(1+p)^n} = 0$.

Special Sequences • 171

Hey, this is the sequence I used to try to make a joke at the end of Chapter 1. I guess it wasn't very funny.... Well, it is now!

This sequence might look a little random, but we will see one of its many applications in the next theorem. Note that it intuitively looks like it converges, since the denominator is "growing" faster than the numerator; in general, exponential growth (expressions like 2^n) get big much faster than polynomial growth (expressions like n^2).

Proof. If we can show that for some constants $b, c \in \mathbb{R}$,

$$0 < \frac{n^\alpha}{(1+p)^n} < cn^b,$$

then if $b < 0$, Theorem 18.2 implies $cn^b \to 0$, so by the pinching theorem, we will have $\frac{n^\alpha}{(1+p)^n} \to 0$.

The denominator $(1+p)^n$ looks like the perfect thing to apply the binomial theorem to. We hope to come up with $(1+p)^n > \gamma n^\beta$, where γ is some constant and $\beta > \alpha$, so that $\frac{1}{(1+p)^n} < \frac{1}{\gamma} n^{-\beta}$. Then we will have $\frac{n^\alpha}{(1+p)^n} < \frac{1}{\gamma} n^{\alpha-\beta}$, and $\beta > \alpha \implies \alpha - \beta < 0$, which is exactly what we want.

We'll start by proving a general identity for the expression $\binom{n}{k}$.

$$\binom{n}{k} = \frac{n!}{(n-k)!k!}$$

$$= \frac{n(n-1)(n-2)\ldots(n-k+1)(n-k)(n-k-1)\ldots(3)(2)(1)}{(n-k)(n-k-1)(n-k-2)\ldots(3)(2)(1)k!}$$

$$= \frac{n(n-1)(n-2)\ldots(n-k+1)}{k!}$$

$$\geq \frac{(n-k+1)(n-k+1)(n-k+1)\ldots(n-k+1)}{k!}$$

$$= \frac{(n-k+1)^k}{k!}.$$

Notice that if $n > 2k$, we have

$$\frac{n}{2} - k > 0 \implies n - k > \frac{n}{2}$$

$$\implies n - k + 1 > \frac{n}{2}$$

$$\implies (n-k+1)^k > \frac{n^k}{2^k}.$$

Thus for any k with $n > 2k$, we have $\binom{n}{k} > \frac{n^k}{2^k k!}$. This is just a constant multiplied by n^k, which will work for us as long as we specify $k > \alpha$.

Putting it all together, let's fix $k \in \mathbb{N}$ with $k > \alpha$. Then for any $n > 2k$, we have

$$(1+p)^n = \sum_{k=0}^{n} \binom{n}{k} 1^{n-k} p^k \quad \text{(by the binomial theorem)}$$

$$\geq \binom{n}{k} p^k$$

$$> \frac{n^k p^k}{2^k k!}.$$

Thus

$$0 < \frac{n^\alpha}{(1+p)^n} < \frac{2^k k!}{p^k} n^{\alpha-k},$$

as long as $n > 2k$. Since $\alpha - k < 0$, the sequence on the right converges to 0. Then by the pinching theorem, so does $\frac{n^\alpha}{(1+p)^n}$.

(Is the fact that we require $n > 2k$ a problem? Not at all! For any sequence $\{s_n\}$, fix a natural number N. Then if the subsequence $\{s_N, s_{N+1}, s_{N+2}, \ldots\}$ converges, so does $\{s_n\}$; because we require that the elements get closer to the limit as n goes to infinity, it doesn't matter at which step the sequence starts.) □

Theorem 18.6. *(The Sequence x^n)*
If $|x| < 1$, then $\lim_{n \to \infty} x^n = 0$.

The condition $-1 < x < 1$ is absolutely crucial. If $|x| = 1$, then the sequence might converge (if it is $1, 1, 1, \ldots$), or it might diverge (if it is $-1, 1, -1, \ldots$). If $|x| > 1$, then at each step the sequence increases by more than it did at the previous step, so it diverges to infinity. Only when $|x| < 1$ can we be sure that it converges.

Proof. There are three possible cases.

Case 1. $x = 0$. Then the sequence $0, 0, 0, \ldots$ converges to 0.
Case 2. $0 < x < 1$. This is where the $\frac{n^\alpha}{(1+p)^n}$ sequence comes in handy. Let $p = \frac{1}{x} - 1$, so $x = \frac{1}{1+p}$ and $p > 0$ (since $x < 1$). Then we let $\alpha = 0$, and apply Theorem 18.5 to see that

$$\lim_{n \to \infty} x^n = \lim_{n \to \infty} \frac{n^0}{(1+p)^n} = 0.$$

Case 3. $-1 < x < 0$. Here, we cannot apply Theorem 18.5 directly, since if we let $p = \frac{1}{x} - 1$, then p isn't necessarily positive (for example, if $x = -\frac{1}{2}$, then $p = -3 < 0$).

Instead, we should actually start by proving $|x|^n \to 0$. Let $p = \frac{1}{|x|} - 1$, so that $|x| = \frac{1}{1+p}$ and $p > 0$ (since $|x| < 1$). Then we let $\alpha = 0$, and apply Theorem 18.5 to see that

$$\lim_{n \to \infty} |x|^n = \lim_{n \to \infty} \frac{n^0}{(1+p)^n} = 0.$$

We can multiply the sequence $|x|^n$ by the constant -1 to see that $-|x|^n$ also converges to 0. We can now apply the pinching theorem to show that $x^n \to 0$, since

$$-|x|^n \leq x^n \leq |x|^n.$$

$\square$

Given these theorems, you should be able to find the limit of almost any convergent sequence you encounter in math or science. Remember, when in doubt, try using the pinching theorem and/or the binomial theorem.

Coming up next, we'll introduce infinite series. It turns out that a series is actually just a certain type of sequence! We both know how much you love sequences by now...

CHAPTER 19

Series

Just like a TV series, a mathematical series can be comedic, dramatic, tragic—or yes, even a soap opera. Series come in all shapes and sizes and can sometimes behave in surprising, nonintuitive ways. In fact, the computation of integrals often uses series, and as such, many people consider series to be the bread-and-butter of real analysis (but I say those people need to start eating a more balanced diet).

In Chapter 2, we briefly mentioned that a series is a sum over all the elements of a sequence. But sequences are infinite, and the concept of an infinite sum is problematic; after all, what does it mean to sum an infinite number of elements? No matter how small each element is, if we sum an *infinite* number of them, won't the result always be infinity?

The answer is no.—Just as with sequences, series can also converge to a limit—but to see how this can happen, we must define series more precisely. It turns out that a series is actually a sequence of sums.

For simplicity, we will limit the definition of series to only real and complex numbers. Of course, we could have series of vectors in $\mathbb{R}^k$, or of elements from any metric space. But why make life so complicated?

Definition 19.1. *(Series)*
For any sequence $\{a_n\}$ in $\mathbb{R}^2$, we define the **partial sum** s_n as

$$s_n = \sum_{k=1}^{n} a_k = a_1 + a_2 + a_3 + \ldots + a_n.$$

*The sequence of partial sums $\{s_n\}$ is called an **infinite series** or simply a **series**. It technically should be written as $s_1, s_2, s_3, \ldots$, which is*

$$\{s_n\} = \{a_1, a_1 + a_2, a_1 + a_2 + a_3, \ldots\}.$$

But for conciseness, we often write

$$\{s_n\} = \sum_{n=1}^{\infty} a_n = a_1 + a_2 + a_3 + \ldots$$

*If $\{s_n\}$ converges to some point $s \in \mathbb{R}$ or $s \in \mathbb{C}$, then we say the series **converges**, and we write*

$$\sum_{n=1}^{\infty} a_n = s.$$

*If no such s exists, then the series **diverges**.*
*The elements of the sequence $\{a_n\}$ are the **terms** of the series.*

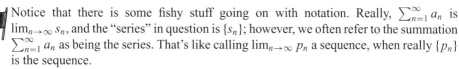Notice that there is some fishy stuff going on with notation. Really, $\sum_{n=1}^{\infty} a_n$ is $\lim_{n \to \infty} s_n$, and the "series" in question is $\{s_n\}$; however, we often refer to the summation $\sum_{n=1}^{\infty} a_n$ as being the series. That's like calling $\lim_{n \to \infty} p_n$ a sequence, when really $\{p_n\}$ is the sequence.

Well, what can I tell you? It's just an imprecise mathematical convention. So when you see "the series $\sum_{n=1}^{\infty} a_n$ converges to s," you should really read

$$s = \lim_{n \to \infty} s_n = \lim_{n \to \infty} \sum_{k=1}^{n} a_k.$$

Always always *always* remember that a series is *not* a sum. A series is a sequence of elements, each of which is a sum. The series converges if and only if this sequence of sums converges.

Because series are just sequences in disguise, every theorem we've proved about sequences also holds for series! No way!

Then why bother studying series in detail, if we already know how to manipulate them? It turns out that they are useful for two reasons:

1. There are certain theorems that work for series in particular, and not for sequences in general (such as the comparison test).
2. There are special series that have many applications, and are thus useful to study (but we need some of those series-specific theorems to prove that they converge).

Sometimes you might encounter the expression $\sum_{n=0}^{\infty} a_n$, instead of $\sum_{n=1}^{\infty} a_n$ (notice where n starts from). Don't freak out! All this means is that the series is the sequence of partial sums over the sequence $\{a_n\}$, but instead of starting from a_1, $\{a_n\}$ happens to start from a_0. Of course $\{a_n\}$ is still a perfectly valid sequence, since there does exist a one-to-one mapping from $\mathbb{N}$ to $\{a_n\}$: $1 \to a_0, 2 \to a_1, 3 \to a_2, \ldots$

Sometimes, when it's obvious where the summation starts and ends, we might just write $\sum a_n$ for short.

One way to check if a series converges is to apply our standard techniques to the sequence of partial sums. But there must be an easier way, right?

Theorem 19.2. *(Convergence of Series)*
A series $\sum a_n$ converges if and only if for every $\epsilon > 0$, $\left|\sum_{k=n}^{m} a_k\right| \leq \epsilon$ for every n and m, both greater than or equal to some natural number N (with $m \geq n$ of course, so the summation is valid).

In symbols, $\sum a_n$ converges if and only if:

$$\forall \epsilon > 0, \exists N \in \mathbb{N} \text{ such that } m \geq n \geq N \implies \left|\sum_{k=n}^{m} a_k\right| \leq \epsilon.$$

The sum inside the absolute value sign is a finite sum, not a series. Notice that it is just the partial sum $s_m = a_1 + a_2 + \ldots + a_{n-1} + a_n + \ldots + a_m$ minus the partial sum $s_{n-1} = a_1 + a_2 + \ldots + a_{n-1}$. So the statement $\left|\sum_{k=n}^{m} a_k\right| \leq \epsilon$ is $|s_m - s_{n-1}| \leq \epsilon$, which is very reminiscent of Cauchy sequences.

Remember that by Theorem 16.10, all Cauchy sequences in Euclidean spaces converge. Since we defined series as sequences in $\mathbb{R}$ or $\mathbb{C}$, we know that any series is Cauchy if and only if it converges. This should make the proof a breeze!

Proof. If $\sum a_n$ converges, then the sequence of its partial sums $\{s_n\}$ converges, so by Theorem 16.3, $\{s_n\}$ is Cauchy. Then given $\epsilon > 0$, there exists an $N - 1$ such that

$$m \geq n \geq N - 1 \implies d(s_m, s_n) < \epsilon, \text{ so}$$

$$m \geq n \geq N \implies m \geq n - 1 \geq N - 1$$

$$\implies |s_m - s_{n-1}| < \epsilon$$

$$\implies \left|\sum_{k=n}^{m} a_k\right| < \epsilon$$

$$\implies \left|\sum_{k=n}^{m} a_k\right| \leq \epsilon.$$

To prove the converse, assume that for every $\epsilon > 0$, there exists an $N \in \mathbb{N}$ such that $\left|\sum_{k=n}^{m} a_k\right| \leq \frac{\epsilon}{2}$ whenever $m \geq n \geq N$. Then given $\epsilon > 0$, there exists an N such that whenever $m \geq n \geq N$, either $m = n$, in which case $d(s_m, s_n) = 0 < \epsilon$, or else we have

$$m \geq n + 1 \geq N \implies \left|\sum_{k=n+1}^{m} a_k\right| \leq \frac{\epsilon}{2}$$

$$\implies \left|\sum_{k=n+1}^{m} a_k\right| < \epsilon$$

$$\implies |s_m - s_{n+1-1}| < \epsilon$$

$$\implies d(s_m, s_n) < \epsilon.$$

Thus the sequence $\{s_n\}$ is Cauchy, and since it is a sequence of complex numbers, Theorem 16.10 implies that it converges. $\square$

This leads us to the following corollary, which is a necessary—but not sufficient—condition for the convergence of a series.

Corollary 19.3. *(Convergence of the Terms of a Series)*
If a series $\sum a_n$ converges, then $\lim_{n \to \infty} a_n = 0$.

Proof. By Theorem 19.2, for every $\epsilon > 0$ there exists an $N \in \mathbb{N}$ such that $|s_m - s_{n-1}| \leq \frac{\epsilon}{2}$ whenever $m \geq n \geq N$. Let $m = n$, so that

$$n \geq N \implies \epsilon > \frac{\epsilon}{2} \geq |s_n - s_{n-1}| = |a_n| = d(a_n, 0).$$

Since this is true for every $\epsilon > 0$, we have $a_n \to 0$. □

The contrapositive is particularly useful. If we have a series $\sum a_n$, and $\{a_n\}$ does not converge to 0, then we know $\sum a_n$ must diverge.

Note that the converse of this corollary is not necessarily true. For example, we know $\frac{1}{n} \to 0$, but as we will learn later (in Theorem 19.10), the series $\sum \frac{1}{n}$ actually diverges. Here is an *incorrect* proof that $a_n \to 0 \implies \sum a_n$ converges. See if you can spot the error!

If $a_n \to 0$, then for every $\epsilon > 0$ there exists an $N \in \mathbb{N}$ such that $|a_n| < \epsilon$ whenever $n \geq N$. Now let $s = \left|\sum_{n=1}^{N} a_n\right|$, so that

$$\left|\sum_{n=1}^{\infty} a_n\right| = \left|\sum_{n=1}^{N} a_n + \sum_{n=N}^{\infty} a_n\right|$$
$$\leq s + \left|\sum_{n=N}^{\infty} a_n\right|$$
$$< s + \epsilon.$$

Thus $\epsilon > \left|\sum_{n=1}^{\infty} a_n\right| - s \geq \left|\sum_{n=1}^{\infty} a_n\right|$, and because this is true for every $\epsilon > 0$, we have $\sum a_n \to s$.

Actually, there is more than one error. First, it doesn't make sense to "split up" the series $\sum_{n=1}^{\infty} a_n$ into two summations $\sum_{n=1}^{N} a_n$ and $\sum_{n=N}^{\infty} a_n$. The symbol $\sum_{n=1}^{\infty} a_n$ (or $\sum a_n$ for short) is just notation that represents the sequence of partial sums $\{s_n\}$. So $\sum_{n=1}^{N} a_n = s_N$, while $\sum_{n=N}^{\infty} a_n$ doesn't mean anything (unless we already know that $\{s_n\}$ converges to s, in which case we would be allowed to write $\sum_{n=N}^{\infty} a_n = s - s_N$, which isn't even helpful here).

Second, in our calculation, we applied the fact that $\left|\sum_{n=N}^{\infty} a_n\right| < \epsilon$. We actually have no idea if this is true! All we know is that $a_n < \epsilon$ for $n \geq N$, but that doesn't mean $\epsilon + \epsilon + \epsilon + \ldots < \epsilon$. That is so ridiculously untrue, it makes me mad. And it should make you mad, too!

Theorem 19.4. *(Bounded Nonnegative Series)*
If a series $\sum a_n$ consists entirely of nonnegative terms ($a_n \geq 0, \forall n \in \mathbb{N}$), then $\sum a_n$ converges if and only if its sequence $\{s_n\}$ of partial sums is bounded.

Of course, the word *nonnegative* restricts this theorem so that it only applies to series in $\mathbb{R}$. Remember that in $\mathbb{C}$ we have no notion of positive and negative, since no order can be defined on the complex numbers.

Proof. Because $\sum a_n$ is a sequence of partial sums $\{s_n\} = a_1, a_1 + a_2, a_1 + a_2 + a_3, \ldots$ in which each term a_n is ≥ 0, the series $\{s_n\}$ is actually a monotonically increasing sequence. Then by Theorem 16.16, the series converges if and only if $\{s_n\}$ is bounded. □

In most cases, we are able to determine the convergence of a series by applying the comparison tests, which are sort of like series' response to sequences' pinching theorem but even better.

Theorem 19.5. *(The Comparison Test for Convergence)*
If there exists an $N_0 \in \mathbb{N}$ such that $|a_n| \leq c_n$ whenever $n \geq N_0$, then if the series $\sum c_n$ converges, so does the series $\sum a_n$.

Proof. If $\sum c_n$ converges, then for any $\epsilon > 0$, by Theorem 19.2 there exists an $N \in \mathbb{N}$ such that

$$m \geq n \geq \max\{N, N_0\} \implies \left|\sum_{k=n}^{m} c_k\right| \leq \epsilon.$$

Thus

$$\left|\sum_{k=n}^{m} a_k\right| \leq \sum_{k=n}^{m} |a_k| \quad \text{(by the triangle inequality)}$$

$$\leq \sum_{k=n}^{m} c_k \quad \text{(since } n \geq N_0\text{)}$$

$$= \left|\sum_{k=n}^{m} c_k\right| \quad \text{(since } 0 \leq |a_n| \leq c_n \text{ for } n \geq N_0\text{)}$$

$$\leq \epsilon.$$

Since this is true for every $\epsilon > 0$, Theorem 19.2 implies that $\sum a_n$ converges. □

Theorem 19.6. *(The Comparison Test for Divergence)*
If there exists an $N_0 \in \mathbb{N}$ such that $a_n \geq d_n \geq 0$ whenever $n \geq N_0$, then if the series $\sum d_n$ diverges, so does the series $\sum a_n$.

Notice that the comparison test for divergence only works when comparing series of nonnegative terms. Because if, for example, each term of some series is $\geq d_n = -1$, it doesn't mean the series diverges just because $\sum d_n = (-1) + (-1) + (-1) + \ldots$ diverges.

Proof. We'll do a proof by contrapositive, so we assume $\sum a_n$ converges and try to show that $\sum d_n$ converges. We can do this in two different (and equally simple) ways:
  Since $|d_n| = d_n \leq a_n$ for every $n \geq N_0$, if $\sum a_n$ converges, then so does $\sum d_n$, by the comparison test for convergence. (Note that we used the fact that $d_n \geq 0$, to see that $|d_n| \leq a_n$.)

Or, if you prefer, we can just apply Theorem 19.4. If $\sum a_n$ converges, its sequence of partial sums must be bounded. Then every partial sum of $\sum d_n$ has an upper bound, so $\sum d_n$ also converges. (Note that here, too, we used the fact that $d_n \geq 0$, since Theorem 19.4 only works on series of nonnegative terms. Actually, we only know that $d_n \geq 0$ for $n \geq N_0$, not for every $n \in \mathbb{N}$; but that's good enough, since if $\sum_{n=N_0}^{\infty} d_n$ converges so does $\sum_{n=0}^{\infty} d_n$, since $\sum_{n=0}^{\infty} d_n$ is just a finite sum added to $\sum_{n=N_0}^{\infty} d_n$.) □

To get the most use out of the comparison tests, we should build up a knowledge base of simple series that converge and diverge so we can frequently compare other series to them.

Theorem 19.7. *(Geometric Series)*
The series $\sum_{n=0}^{\infty} x^n$ converges to $\frac{1}{1-x}$ if $|x| < 1$, and diverges if $|x| \geq 1$.

Any such series of terms raised to the *n*th power is called a *geometric series*. These geometric series pop up all over math, and this formula will come in handy.

Proof. The divergence case of $x \geq 1$ is the easy part. We see that $\sum x^n$ diverges by the comparison test, since $x^n \geq 1$ for every $n \in \mathbb{N}$, and the series $1 + 1 + 1 + \ldots$ clearly diverges.

If $x \leq -1$, then $|x^n| \geq 1$ for every $n \in \mathbb{N}$. Then $|x^n - 0| \geq 1$ for every $n \in \mathbb{N}$, so it can't be true that $x_n \to 0$. Thus by the contrapositive of Corollary 19.3, the series diverges.

For convergence, let's start by figuring out a formula for the partial sum

$$s_n = \sum_{k=0}^{n} x^k = 1 + x + x^2 + \ldots + x^n,$$

since the limit of the series $\sum x^n$ is the limit of the sequence $\{s_n\}$. By some algebraic manipulation, we get

$$(1-x)s_n = (s_n - xs_n)$$
$$= (1 + x + x^2 + \ldots x^n) - (x + x^2 + x^3 + \ldots + x^{n+1})$$
$$= 1 + (x - x) + (x^2 - x^2) + \ldots + (x^n - x^n) - x^{n+1}$$
$$= 1 - x^{n+1},$$

so $s_n = \frac{1-x^{n+1}}{1-x}$. By Theorem 18.6, we know $x^n \to 0$ when $|x| < 1$, so that $\frac{1-x \cdot x^n}{1-x} \to \frac{1}{1-x}$. Thus

$$\sum_{n=0}^{\infty} x^n = \lim_{n \to \infty} s_n = \frac{1}{1-x} \quad (|x| < 1).$$

□

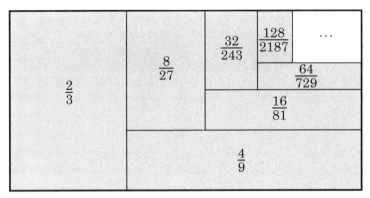

Figure 19.1. Filling in a 1×2 rectangle with the series $\sum_{n=1}^{\infty} \left(\frac{2}{3}\right)^n$.

Note that if we want to start the series at $n = 1$ (instead of $n = 0$), we have

$$\sum_{n=1}^{\infty} x^n = \sum_{n=0}^{\infty} x^n - x^0 = \frac{1}{1-x} - 1 = \frac{x}{1-x}.$$

Example 19.8. (Geometric Series)
Geometric series have an especially nice geometric representation (coincidence?!). Let's use the example of $x = \frac{2}{3}$, for which our formula tells us that

$$\sum_{n=1}^{\infty} \left(\frac{2}{3}\right)^n = \frac{\frac{2}{3}}{1 - \frac{2}{3}} = 2.$$

So if we start with a rectangle of area 2, we should be able to "fill it in" using this series, as we can see in Figure 19.1.

Take a 1×2 rectangle, and fill in an area of size $\frac{2}{3}$. The portion that remains has area $2 - \frac{2}{3} = \frac{4}{3}$.

Next, we add $\left(\frac{2}{3}\right)^2 = \frac{4}{9}$ to our shaded area; notice that $\frac{4}{9}$ is exactly one third of the area that remains. Now the portion that remains has area $\frac{4}{3} - \frac{4}{9} = \frac{8}{9}$.

Next, we add $\left(\frac{2}{3}\right)^3 = \frac{8}{27}$ to our shaded area, which is exactly one third of the area that remains. Now the portion that remains has area $\frac{8}{9} - \frac{8}{27} = \frac{16}{27}$.

Basically, at each step we fill in one third of the remaining area. If we continue this pattern forever, we will have filled in the entire rectangle of area 2.

Before seeing another commonly used series, we need to spruce up our toolbox with one more test for series convergence.

Theorem 19.9. *(The Cauchy Condensation Test)*
A series $\sum_{n=1}^{\infty} a_n$ of nonnegative, monotonically decreasing terms converges if and only if the series $\sum_{k=0}^{\infty} 2^k a_{2^k}$ converges.

So from now on, whenever we have a series built from terms $\{a_n\}$ with $a_1 \geq a_2 \geq a_3 \geq \ldots \geq 0$, we can use the Cauchy condensation test to check for convergence. It's pretty cool that a series containing so few elements of $\{a_n\}$ (namely, $a_1, a_2, a_4, a_8, \ldots$) still determines the convergence or divergence of $\sum a_n$!

Proof. Because $\sum_{n=1}^{\infty} a_n$ and $\sum_{k=0}^{\infty} 2^k a_{2^k}$ are series of nonnegative terms, by Theorem 19.4, each one converges if and only if its sequence of partial sums is bounded. Set

$$s_n = a_1 + a_2 + a_3 + \ldots + a_n \text{ (the } n\text{th partial sum of } \sum_{n=1}^{\infty} a_n),$$

$$t_k = a_1 + 2a_2 + 4a_4 + \ldots + 2^k a_{2^k} \text{ (the } k\text{th partial sum of } \sum_{k=0}^{\infty} 2^k a_{2^k}).$$

So if we can show $\{s_n\}$ bounded $\iff \{t_k\}$ bounded, we will have proven $\sum_{n=1}^{\infty} a_n$ converges $\iff \sum_{k=0}^{\infty} 2^k a_{2^k}$ converges. Let's prove both directions of the "iff" implication.

1. $\{t_k\}$ bounded $\implies \{s_n\}$ bounded. We already know that $\{s_n\}$ and $\{t_k\}$ are bounded below (by 0), so by Theorem 9.6 we just need to worry about the upper bound. Assume there is an $M \in \mathbb{R}$ with $t_k \leq M$ for every $k \in \mathbb{N}$; if we can show that for every $n \in \mathbb{N}$ there is a $k \in \mathbb{N}$ such that $s_n \leq t_k$, we will have shown that every element of $\{s_n\}$ is also $\leq M$.

   Well, that's not so hard! If we fix n and choose a k such that $n < 2^k$, then we have

$$s_n = a_1 + a_2 + a_3 + \ldots + a_n$$
$$\leq a_1 + a_2 + a_3 + \ldots + a_{2^{k+1}-1} \quad \text{(since } n < 2^k \implies n + 1 < 2(2^k) - 1\text{)}$$
$$= a_1 + (a_2 + a_3) + (a_4 + a_5 + a_6 + a_7) + \ldots + (a_{2^k} + \ldots + a_{2^{k+1}-1})$$
$$\leq a_1 + 2a_2 + 4a_4 + \ldots + 2^k a_{2^k} \quad \text{(since } a_2 \geq a_3 \geq a_4 \geq \ldots\text{)}$$
$$= t_k.$$

   Thus for every s_n, we have

$$s_n \leq t_{\left\lceil \frac{\log(n)}{\log(2)} \right\rceil + 1} \leq M,$$

   so $\{s_n\}$ is bounded. (Note that we chose $k = \left\lceil \frac{\log(n)}{\log(2)} \right\rceil + 1$ so that $2^k > n$.)

2. $\{t_k\}$ unbounded $\implies \{s_n\}$ unbounded. Assume that for every $M \in \mathbb{R}$ there is a $k \in \mathbb{N}$ such that $t_k > M$; we want to show that for every $N \in \mathbb{R}$, there is an $n \in \mathbb{N}$ such that $s_n > N$.

   Now we do the opposite of the previous direction: take any k, and choose and n such that $n > 2^k$, so we have

$$s_n = a_1 + a_2 + a_3 + \ldots + a_n$$
$$\geq a_1 + a_2 + a_3 + \ldots + a_{2^k} \quad \text{(since } n > 2^k\text{)}$$
$$= a_1 + a_2 + (a_3 + a_4) + (a_5 + a_6 + a_7 + a_8) + \ldots + (a_{2^{k-1}+1} + \ldots + a_{2^k})$$
$$\geq \frac{1}{2}a_1 + a_2 + 2a_4 + 4a_8 + \ldots + 2^{k-1} a_{2^k} \quad \text{(since } a_2 \geq a_3 \geq a_4 \geq \ldots\text{)}$$
$$= \frac{1}{2}(a_1 + 2a_2 + 4a_4 + \ldots + 2^k a_{2^k})$$
$$= \frac{1}{2}t_k.$$

Then for every $N \in \mathbb{N}$, let $M = 2N$. Thus there is a $k \in \mathbb{N}$ with $t_k > M$, so we have

$$s_{2^k+1} \geq \frac{1}{2}t_k > \frac{1}{2}(2N) = N,$$

so $\{s_n\}$ is unbounded. (Note that we chose $n = 2^k + 1$ so that $n > 2^k$.) □

One of the best applications of the Cauchy condensation test is for determining the convergence of *p-series*, which are of the form $\sum \frac{1}{n^p}$.

The most famous such series is $\sum \frac{1}{n}$, which is called the *harmonic series*. The name comes from harmonic series in music, which are notes whose sound waves have $\frac{1}{2}, \frac{1}{3}, \frac{1}{4}, \ldots$ the wavelength of the starting note.

Theorem 19.10. *(p-Series)*
The series $\sum_{n=1}^{\infty} \frac{1}{n^p}$ converges if $p > 1$, and diverges if $p \leq 1$.

Note that this implies that the harmonic series $\sum \frac{1}{n}$ diverges.

Proof. If $p < 0$, then the sequence $\{n^{-p}\}$ is unbounded, so it does not converge. Then by the contrapositive of Corollary 19.3, $\sum \frac{1}{n^p}$ must also diverge.

If $p \geq 0$, then every element of the sequence $\{\frac{1}{n^p}\}$ is less than or equal to the previous element (and of course all of them are positive), so we can apply the Cauchy condensation test. It turns out that the series

$$\sum_{k=0}^{\infty} 2^k \left(\frac{1}{(2^k)^p}\right) = \sum_{k=0}^{\infty} (2^{1-p})^k,$$

is just a geometric series (where $x = 2^{1-p}$). By Theorem 19.7, it converges if $0 \leq 2^{1-p} < 1$ (so if $1 - p < 0$) and diverges if $2^{1-p} \geq 1$ (so if $1 - p \geq 0$). Thus $\sum \frac{1}{n^p}$ converges if $p > 1$, and diverges if $p \leq 1$. □

With the tools from this chapter, you should be able to deal with a variety of series. The comparison tests (usually comparing to a geometric series or a *p*-series) and the Cauchy condensation test are only the beginning of the complete picture; there's also the root test and the ratio test.

So why care about series? Because they care about you. Also because they are used in the definition of the number e, the complex analysis definition of π—and if you've studied Taylor series, you've seen how almost any function can be written as a series! Series give you a better understanding of convergent sequences and the meaning of infinity.

CHAPTER 20

Conclusion

We've covered a lot of material, which is both a blessing and a curse. The blessing is that you are now familiar with many important topics, which you will cherish for the rest of your life (I hope). The curse is that when faced with new problems on a homework assignment or a test, the "use all available information" advice sounds ridiculous. There are so many facts to remember! That's why working backward on the problem—starting with what you need to prove—can be so helpful.

As an additional aid, here is a compilation of our greatest hits: the best tricks we learned from each chapter that we continued to use many times later on.

Chapter 1. When reading anything math-related, read actively! Go slowly and take notes.

Chapter 2. Sometimes a seemingly complicated proof can be accomplished by a simple method: counterexample, contrapositive, contradiction, or induction. (You can review Examples 2.1, 2.2, 2.3, and 2.4.)

Chapter 3. To prove $A = B$ for two sets, just show $A \subset B$ and $B \subset A$. (You can review the proof of Theorem 3.12.) Also, remember to use De Morgan's law, which says that the complement of the union is the intersection of the complements (see Theorem 3.17).

Chapter 4. Use both properties of a least upper bound: no number in the set can be higher than it, and anything lower than it isn't an upper bound of the set. (You can review the proof of Theorem 4.9.)

Chapter 5. Use the Archimedean property, which tells you that for any real x and y there's an $n \in \mathbb{N}$ with $nx > y$ (see Theorem 5.5). In it's simplest form, it says there's always an $n > y$.

Chapter 6. Use the triangle inequality (see Property 5 of Theorem 6.7), or its big brother, the Cauchy-Schwarz inequality (see Theorem 6.8).

Chapter 7. Use all three properties of a bijection: it is a function (so it is defined for the whole domain, and no element maps to two different elements), it is injective (so no two elements map to the same element), and it is surjective (so every element of the codomain is mapped to). (You can review the proof of Theorem 7.16. Hopefully you filled in those blanks!)

Chapter 8. To prove that a set is countable, if you found a likely bijection but it has the pesky problem of possibly containing duplicates, define a "bijective version" of that function that acts on a subset of the domain. (You can review the proof of Theorem 8.16.)

Chapter 9. You're often trying to construct a neighborhood to satisfy a condition (that it's contained inside a set, or it contains another point, etc.), so work backward to figure out the "magic radius" for your neighborhood. (You can review the proof of Theorem 9.23.)

Chapter 10. Sometimes it's easier to work with the complement of a set rather than the original set. When you do, openness and closedness get reversed. (You can review the proof of Theorem 10.7.)

Chapter 11. Use the Russian doll property, also known as the fact that the infinite intersection of nested compact sets is nonempty (see Corollary 11.12).

Chapter 12. The Heine-Borel theorem tells us that in $\mathbb{R}^k$, being compact is equivalent to being closed and bounded (see Theorem 12.6).

Chapter 13. Many times you can break down a topology problem into two simple cases: if $p \in A$, what does that imply? And if $p \notin A$, what does that imply? (You can review the proof of Theorem 13.8.)

Chapter 14. If you know a series converges, the fact that $n \geq N \implies d(p_n, p) < \epsilon$ for all $\epsilon > 0$ means that any ϵ works, including numbers like $\frac{\epsilon}{2}$. (You can review the proof of Theorem 14.6.)

Chapter 15. You can construct a specific subsequence similar to doing a proof by induction: define p_1, then assume something holds true for p_{n-1} and show it's also true for p_n. Don't forget to first address the possible case where the range of $\{p_n\}$ is finite. (You can review the proof of Theorem 15.7.)

Chapter 16. In $\mathbb{R}^k$ (or any complete metric space), showing that a sequence is Cauchy is enough to prove that it converges (see Theorem 16.10).

Chapter 17. Don't be afraid to take the subsequence of a subsequence! This is especially useful when you can show that a subsequence either has a divergent subsequence, or else is bounded so it has a convergent subsequence. (You can review the proof of Theorem 17.8.)

Chapter 18. Use the pinching theorem (see Theorem 18.1). Also, when working with exponents, you'll often have to use the binomial theorem $(a+b)^n = \sum_{k=0}^{n} \binom{n}{k} a^{n-k} b^k$ to expand the power, and then you can drop off most of the terms when showing it is $\geq$ or $>$ something. (You can review the proof of Theorem 18.3.)

Chapter 19. Series are *not* sums, they are sequences! So you can apply everything you know about sequences to series, plus you have the comparison tests (see Theorems 19.5 and 19.6) and the Cauchy condensation test (see Theorem 19.9).

Of course, this is only a small subset of your toolbox. Whenever you're stuck on a proof, try working backward: narrow down what you have to prove until it's an obvious statement. Make a list of the assumptions used in the theorem, and try using all of them. Work with the simplest examples before you try to do the general case. And *draw pictures.* Seriously! The more pictures in your notebook, the better you are (as a student, and as a human being).

It's possible that your course covers more material beyond the topic of series. Although you might be intimidated by what lies ahead in your grand real analysis journey—continuity, derivatives, integrals, sequences of *functions* (what kind of horror is *that?!*)—remember that it all comes back to the basics. Everything relies on real numbers, topology, and sequences. If you can master those topics, the rest will be a breeze! Really.

Also, for every minute you've spent trying to understand these concepts, you've probably spent at least that much time figuring out how proofs work. Hopefully by now, you've mastered the basic techniques of proofs—reading them and writing them—and will soon be able to focus purely on the math.

My hope is that you learned more from this book than just how to prove that a sequence converges (you *did* learn that, right?). You learned how to think rigorously; you learned how to deal with infinity, in cases where your intuition wouldn't help (for example, with countability or series). Most of all, I hope you learned that going slowly and understanding the definitions first can make all the difference.

Have fun! It may seem scary, but real analysis can be incredibly interesting and exciting. You can do it.

ACKNOWLEDGMENTS

First, a big thanks to you, the student, for reading the acknowledgments. ...But seriously, why are you reading this? Get back to math!

This book began as my thesis project at Princeton University. My heartfelt gratitude goes out to Dr. Philippe Trinh, my ever-helpful and enthusiastic adviser. He taught me the value of good style, good formatting, and good pictures. From the beginning, he believed in this project with a passion to rival my own—so everything here is thanks to him.

Thank you to Adrian Banner, my second adviser, for his help and for inspiring this textbook series with the incredible *The Calculus Lifesaver*.

Thank you to Mark McConnell, for his detailed editing and insightful feedback.

Thank you to the staff at Princeton University Press, especially my editor Vickie Kearn, who supported me early on and worked hard to make this book happen.

Thank you to the generous users of tex.stackexchange.com, who constantly volunteer their own time to help with the LaTeX problems of complete strangers.

To my wife Charlotte, who is the sup to my inf, the Heine to my Borel, the N to my ϵ: thank you for editing this book (or at least, just the acknowledgments section). I love you.

To my family Monica, Joel, Joshi, Greg, Jacquy, Eliane, Agar, Paul, and Marnie: thank you for supporting me during the writing of this book—and during my life in general.

And to the many wonderful teachers I've had over the years: thank you for inspiring me to love learning almost as much as to love teaching.

BIBLIOGRAPHY

1. **Walter Rudin**, *Principles of Mathematical Analysis*, 3rd edition (McGraw-Hill, 1976). This book follows the exact curriculum of Rudin's. Most definitions and proofs here are due to him.

2. **Steven R. Lay**, *Analysis with an Introduction to Proof*, 4th edition (Prentice Hall, 2004). This one gets my highest recommendation. Nice explanations; lots of preliminary chapters that introduce you to logic, sets, and functions; and encouragement of active reading via in-chapter practice questions and fill-in-the-blanks (which I obviously love).

3. **Stephen Abbott**, *Understanding Analysis* (Springer, 2010).
   With prechapter "Discussions" and postchapter "Epilogues," this book has a nice structure with lots of new topics and examples to enrich the traditional Rudin-based curriculum. What's best about this book is Abbott's efforts to motivate the question "why study real analysis?" with introductory problems for which your previous math intuition provides no help. He places emphasis on constructing a unifying narrative to connect all the chapters. (Of course, this narrative and historical approach is only effective when you already know the material and you are interested in going back and learning what it all means, how it came about, and why it's all related. For the first-time student of analysis, understanding abstract definitions and writing rigorous proofs is enough of a challenge, without the added burden of trying to put it all together in your head.)

4. **Kenneth A. Ross**, *Elementary Analysis: The Theory of Calculus* (Springer, 2010).
   Ross clearly gives up quantity in favor of quality. His pages are full of detailed examples and clear explanations, although there are some holes in the material he covers, and his book falls short of the scope of some first-semester analysis courses. Definitely read his second chapter if you are having any trouble with sequences (especially proving that sequences converge and understanding lim sups).

5. **Robert G. Bartle** and **Donald R. Sherbert**, *Introduction to Real Analysis*,
   3rd edition (Wiley, 2000).
   This book is comprehensive and detail-oriented. It covers most of Rudin (in a somewhat jumbled order) and fills in many missing bits of theorems and examples that could be helpful. The explanations of difficult concepts are a mixed bag, but you should take a look if you need help with proofs by induction or a review of functions.

6. **Robert Wrede** and **Murray Spiegel**, *Schaum's Outline of Advanced Calculus*,
   3rd edition (McGraw-Hill, 2010).
   Schaum's is a large set of practice exercises, some of which have detailed solutions. Solving problems is better than reading about them, so you should check this one out.

7. **Burt G. Wachsmuth**, *Interactive Real Analysis*, version 2.0.1(*a*)
   (www.mathcs.org/analysis/reals, 2013).
   This website provides excellent interactive examples and proofs in which you can think about them first, then click to see the answers.

INDEX

! (factorial), 169
(a, b) (open interval), 16
$<$ (less than), 29
$>$ (greater than), 29
E' (limit points), 93
$N_r(p)$ (neighborhood), 82
$[a, b]$ (closed interval), 16
$|\ |$ (absolute value), 50
$\binom{n}{k}$ (n choose k), 169
$\perp$ (contradiction), 38
$\cap$ (intersection), 17
$\lceil\ \rceil$ (ceiling), 32
$\circ$ (composition), 67
: (mapping), 61
$\cup$ (union), 17
$\emptyset$ (empty set), 15
$\equiv$ (equivalence relation), 68
$\exists$ (exists), 6
$\forall$ (for all), 6
$\geq$ (greater than or equal), 29
$\iff$ (if and only if), 7
$\implies$ (implies), 7
$\in$ (in), 14
∞ (infinity), 45
$\leq$ (less than or equal), 29
$\lim_{n\to\infty}$ (limit), 130
$\liminf_{n\to\infty}$ (lower limit), 159
$\limsup_{n\to\infty}$ (upper limit), 159
$\mapsto$ (maps), 61
$\mathbb{C}$ (complex numbers), 46
$\mathbb{N}$ (natural numbers), 7
$\mathbb{N}_n$ (proper subset of $\mathbb{N}$), 70
$\mathbb{Q}$ (rational numbers), 7
$\mathbb{R}$ (real numbers), 36
$\mathbb{R}$ (real numbers), 7
$\mathbb{R}^2$ (2-vectors), 46
$\mathbb{R}^k$ (Euclidean space), 54
$\mathbb{Z}$ (integers), 7
$\neg$ (not), 8
$\overline{E}$ (closure), 93
$\overline{z}$ (conjugate), 49
$\setminus$ (complement), 21
$\sim$ (cardinality), 68
$\sqrt[n]{\ }$ (nth root), 40

$\square$ (Q.E.D.), 13
$\subset$ (subset), 16
$\subseteq$ (subset), 15
$\sum$ (sum), 7
$\sum_{n=1}^{\infty}$ (series), 7, 174
$\supset$ (superset), 16
$\supseteq$ (superset), 15
$\to$ (limit), 130
$\to$ (maps), 61
^{C} (complement), 21
s^* (upper limit), 159
s_* (lower limit), 159

absolute value, 50
accumulation point, 82
algebra, 35
arbitrarily, 9
Archimedean property, 37
associative; under addition, 35; under multiplication, 35
at most countable, 70
axiom, 4
axiom of completeness, 37

base case, 11
bijection, 66
binomial theorem, 168
Bolzano-Weierstrass, 146
bounded, 80; sequence, 129
bounded above, 30
bounded below, 30

Cantor diagonal process, 73
Cantor set, 120
cardinal number, 68
cardinality, 68
Cauchy condensation test, 180
Cauchy sequence, 148
Cauchy-Schwarz inequality, 52
ceiling, 32
choose, 169
closed, 84; relative, 97; under addition, 35; under multiplication, 35
closed interval, 16